Socio-economic Effects and Constraints in Tropical Forest Management

Socio-economic Effects and Constraints in Tropical Forest Management

The results of an enquiry organized by the
International Federation of Institutes for Advanced Study

Edited by
E. G. Hallsworth
IFIAS
Science Policy Research Unit
University of Sussex
Brighton, UK

A Wiley–Interscience Publication

175 YEARS OF PUBLISHING
1807 1982

JOHN WILEY & SONS

Chichester · New York · Brisbane · Toronto · Singapore

Library of Congress Cataloging in Publication Data:

Main entry under title:

Socio-economic effects and constraints in tropical
forest management.

'A Wiley–Interscience publication.'
Papers presented at the Workshop on Socio-Economic
Effects and Constraints in Forest Management,
held at Dehra Dun, India, 1981.
Includes index.
1. Forest management—Tropics —Congresses.
2. Forests and forestry— Economic aspects—
Tropics—Congresses. 3. Forests and forestry—
Social aspects—Tropics—Congresses. 4. Forest
policy—Tropics—Congresses. I. Hallsworth, E. G.
(Ernest Gordon) II. Workshop on Socio-Economic
Effects and Constraints in Forest Management
(1981 : Dehra Dun, India)
SD247.S63 333.75′0913 82-1989

ISBN 0 471 10375 6 AACR2

British Library Cataloguing in Publication Data:

Socio-economic effects and constraints in
tropical forest management.
1. Forest management—Congresses
I. Hallsworth, E. G.
634.9′28 SD561

ISBN 0 471 10375 6

Typeset by Preface Ltd., Salisbury, Wilts.
Printed in the United States of America

Contents

Foreword

Shri M. K. Dalvi
Inspector General of Forests, India.

It is indeed a great pleasure and honour for me to be here to inaugurate this Workshop on Socio-economic Effects and Constraints in Forest Management. We are meeting at this centre of Indian forestry which, since its inception some seven decades ago, has contributed significantly towards the understanding of the multifarious problems connected with, and related to, forests and forestry. This is also the institute where the foresters in India are educated and trained. It is, therefore, quite befitting that this NORAD/IFIAS/GOI Workshop is being held here.

The subject matter of this Workshop—Socio-economic Effects and Constraints in Forest Management— to my mind, is very important from the point of the developing nations, generally characterized by tropical and semi-tropical vegetations. Let me first of all confess that I am neither a sociologist nor an economist. I am a forester, as you all are aware. You will, therefore, bear with me on my random thoughts on this subject.

While some of the effects and constraints affecting forest management may differ among different countries, others appear to be common to all. This is as expected, since we are primarily talking of people and sylvan surroundings, and the interactions between them as related to forest management.

All practising foresters present here will, I am sure, recall that at some point of time in their career they have been bewildered by the decisions taken against their seemingly sound and scientific advice; the violent reaction to some of the innocent management decisions like the choice of plantation species, or the replacement of an overmature crop. Indignant foresters have considered such reactions as arising out of ignorance, ill-conceived, anti-national, irrational, politically motivated and so on.

These decisions and reactions are basically socio-economic in origin. Since this is a gathering of experts, who will I presume be discussing these factors in greater detail, further elaboration at this stage may not serve any useful purpose. What seems important, however, is the need for identifying the causal factors to assess whether the so-called 'anti-forest' actions stem from the same causes as attributed by the foresters, or whether they do reflect in fact some overlooked aspects basic to forest management.

This Workshop—which is being attended by participants from several nations—will, I hope, enable us to develop a clearer understanding of such overlooked factors—social, psychological, economic, niotic, or whatever they might be—that lead to adverse effects and often act as constraints to forest management. I myself have a few observations to make which I believe will be of some relevance in this context.

Most socio-economic effects and constraints to forest management, in my opinion, have their roots in the behavioural attitudes of the people and fundamental characteristics of the forests. An appreciation of these factors may not only assist in concretizing the issues involved, but may also help the deliberations in devising meaningful forest management practices.

A large number of factors, singly or in various combinations, lead to socio-economic effects and constraints in forest management, among which the more important are valuation of benefits of a particular resource and time preference attached to it by nation, society, various group classes and individuals, lack of appreciation by forest managers of cost and benefits in real terms, differences between financial and economic benefits, general ignorance of people, and the communication gap between the forest managers and the rest.

It would be interesting to consider how some of these factors react and interact leading to socio-economic effects and constraints in forest management.

It is easy to see that, given a resource with its limitations of output of goods and services, every person tries to maximize the consumption option of his choice. Put in the mathematician's jargon this might be labelled the individual's 'objective function'. Similarly, there would be objective functions related to various groups brought together by common interest, society and nation. It is most unlikely that they will all correspond. Let us consider a simple example of a tree growing near a village. This tree may mean different things to different people. To a child it is a convenient place to rig up a swing, to a village matron it is a source of fuelwood, to an industrialist it is a source of raw material, and to a recreationist perhaps it is the very charm of the countryside. Given this particular tree, each would like to maximize the output represented by it. Management practices of the forester could, perhaps, accommodate some of these, but not all, since the uses are not always mutually compatible. Conversion of the tree into industrial raw material may satisfy the industrialist and to some extent the village matron, but will prevent the child and the recreationist from maximizing their individual objective functions. Let us consider another example of a natural forest predominantly of sal. This forest represents to poor forest-fringe-dwellers a source of livelihood yielding seeds for sale, branches and leaves for fuel and manure. The decision to convert this sal forest to industrially more valuable species like teak may satisfy the needs for higher revenues which may or may not be used for the welfare of these same people, but would certainly deprive them of an output from the forest which they were enjoying.

Obviously distribution of welfare is also affected in these and similar cases because of such management decisions where benefits are not shared proportionately among various classes, groups or individuals. The forester's management practices may consequently suffer and may be totally frustrated, creating conditions of abnormal stress and tension. An appreciation of the factors causing stresses and strains may lead a forester to take more appropriate management decisions.

This however poses certain questions before me.

The first is about the present-day forester's education and training. Is he equipped with education and training to appreciate and evaluate the socio-economic problems? Or is he provided with no tools to cope with these problems?

The second question is regarding the attitude of the present-day foresters. Is everything well on this front?

I do not want to dwell on these issues for the fear of pre-empting your discussions, but I am sure this Workshop will provide suitable answers to these and similar questions. I wish this Workshop a great success, and declare it open.

Preface

It is interesting to reflect that in spite of the volume of work published on tropical forests over the last 10 years, relatively little has appeared on the impact that tropical forest management might have on lives of the people who live in and on the fringes of the forests. This may perhaps be related to the observation that the bulk of forestry research, and most studies of forest management practices, have been undertaken in the cool temperate forests of the northern hemisphere in which, in general, men do not live. By contrast most tropical forests are inhabited, and in many parts of the world are inhabited by remnants of tribes for whom the forest was the last refuge into which they retreated before the greater numbers, superior weaponry or better organization of an invading race.

When NORAD, the Norwegian Agency for International Development, asked the International Federation of Institutes of Advanced Study (IFIAS) to organize a seminar in India on tropical forest management, it was logical for IFIAS to accept, for the main purpose of IFIAS is the organization of interdisciplinary studies that will benefit by international comparisons.

The material on which this book is based was derived from the papers presented at the meeting that resulted—a Workshop on the socio-economic effects and constraints of tropical forest management, held at Dehra Dun, India, convened by IFIAS in cooperation with the Indian Forest Research Institute, with the financial support of NORAD and the Government of India.

Not all of the material presented at the working party has been included in this book. Those not included were of value to the working party in the execution of its task of advising the IFIAS Board on the desirability of some future lines of enquiry, but because of the particular aspects that they dealt with, were not appropriate for inclusion here.

The workshop was opened by Shri M. K. Dalvi, Inspector-General of Forests for India. His opening address serves most appropriately as the foreword to this book.

Brighton, October 1981 E. G. HALLSWORTH

Socio-economic Effects and Constraints in Tropical Forest Management
Edited by E. G. Hallsworth

1

The human ecology of tropical forest

E. G. HALLSWORTH
The IFIAS–SOS Project, Science Policy Research Unit, University of Sussex, UK

INTRODUCTION

Soil science and plant ecology have, ever since their inception, embraced two similar fundamental concepts which it is part of the purpose of this book to challenge. These are the concepts of the 'natural' situation and of the 'mature' or 'climax' state.

The criterion of the 'natural' situation is that only the 'natural' features of the environment have played a part in development, and hence that no human activity has disturbed their interplay. It seems doubtful, at this time, whether there is much of the earth's surface outside the boreal and alpine areas where this criterion can apply. It certainly cannot in many and perhaps most tropical forests, which have probably been occupied by man for much of recent time.

It is generally accepted that the nature of the vegetative cover, in each different locality on the earth, is related primarily to the climate and soil that occurs in each locality. A 'climax' vegetation consequently should only develop on a mature soil.

The soil itself, however, is not an independent factor, but reflects the interplay between the climate, the nature of the underlying rock or rock debris on which the soil has formed, the topography of the site and the organisms that are present in and on the soil. As Jenny (1941) pointed out, the system is not static, and the nature of the soil found at any one time in the history of the land surface of a particular area of the world will depend on the length of time that the set of soil-forming factors found at the present have been operating. The rate of change as measured by change in the quantity and distribution of the different soil constituents will vary from time to time, and when the rate of change approaches zero the soil could be considered to be mature.

In any part of the world where for a part of the year rainfall exceeds evaporation, the excess moisture will drain away out of the soil and will carry with it soluble constituents including organic matter and various ions, such as potassium and calcium. This effect has been well demonstrated by the work on the dune sands at Southport, UK (Salisbury, 1925) or the polders in Holland (Hissink, 1938) or on the dune sequence at Woy Woy in New South Wales, Australia (Burges and Drover, 1953). All of these studies show a progressive loss of calcium with time, and a progressive acidification of the topsoil. Vegetation will establish on the soil during this time, and presumably when the rate of change of acidity and the change in rate of release of nutrient ions becomes sufficiently slow the vegetative pattern will stabilize and, in current concepts, this will be the climax vegetation. On soil materials with a high proportion of weatherable minerals this state of minimal change will be long-lasting, but the system is in a meta-stable state, and the 'climax' vegetation will persist only as long as the rate of release of calcium, etc. from the weatherable minerals is equal to their rate of loss in the drainage waters. On soil parent materials low in weatherable minerals, and in areas where the excess of rainfall over evaporation leads to considerable throughput of drainage water, the rate of change in the nature of the soil will be greater and the change in the 'climax' vegetation more easily demonstrated.

Studies on the Cooloola dunes on the sub-tropical coast of Queensland illustrate this effect (Thompson, 1981). Here, in an area where rainfall and temperature regimes are suitable for rainforest, the oldest dunes—now notably lower in calcium and potassium—carry only a heath vegetation, the mid-age dunes carry sclerophyll forest, whilst the younger dunes carry rainforest species. In each case the vegetation found could be considered as the climax vegetation for the soil concerned and the climate of the mid-Queensland coast. Since the pedological studies allow the sites to be ranked in order of age, it would seem logical to consider the vegetation found as a sequence of meta-stable climaxes, of which only the last may be the true climax, because here nutrient loss in the drainage waters might perhaps be balanced by nutrient input in dust and rain.

In a highly leached soil, in a leaching environment, the most successful plant species will be those with a root system which provides a highly efficient mechanism for trapping nutrients, to some extent by deep rooting, but also by developing a root mass which spreads across the surface as do the proteaceous components of the Australian flora (Malajczuk and Bowen, 1974) or many of the species of the forests of Amazonia (see Chapter 14 of this book). In any such system, where the nutrients required by the plants are held in a closed cycle, any operation which leads to a break in the cycle, allowing some part of the nutrients previously held in the cycle to flow out of the system either in drainage waters, into the atmosphere or by export in the form of timber, crops or stock, will have a deleterious effect on the subsequent growth of the plant component of the system.

The organisms present in the soil or on the land surface are not all producers of organic matter, and the quantity of biomass produced in any system, or the quantity of organic matter deposited on and subsequently incorporated into the soil, depends on the types of organisms present and on the balance between the activities of the synthesizer and the decomposer species. The decomposer species include not only the saprophytic bacteria and fungi which are mainly responsible for the breakdown of the plant debris falling on the soil surface; they also include all those other species which derive their energy from the photosynthate of the synthesizer species—the fungal and bacterial pathogens, the whole of the insect fauna which prey on plants, the worms, the rodents and other species normally considered as pests, but also grazing animals such as cattle, deer and elephants etc., and finally, of course, man himself.

Most ecological studies have been attempts to investigate so-called 'natural' systems; that is systems that have developed in the absence of any interference by man. It is open to doubt whether there is much of the land area of the world where the vegetation has been totally unaffected by man. Even in Australia—where until British settlement began at the end of the eighteenth century there were only perhaps some 300,000–350,000 people living in a continent of 769 million hectares, and obtaining their needs only by hunting and gathering—the deliberate use of fire that was practised by the Aborigines had a marked effect on the vegetation of the country. There is now, however, a more widespread appreciation of anthropogenic affects in the tropics, as for example, in the production of the kunai grass or alang-alang areas of the islands of the western Pacific.

In this book an attempt is made to look at the human factor in the ecology of tropical forests; to consider man as one of the decomposer species It is consequently necessary to consider the components of human demand. These include not only the demands of the people who live in the forest, whose inclusion as part of the ecosystem could be accepted with little strain, but also those people who do not live in the forests, but whose demands for timber, for water and for recreation may totally dominate the ultimate pattern of use and management. The manner in which these demands are being met differs in different parts of the world, and to illustrate the problems I asked several of my colleagues from different parts of the world to look at the effects of man's demands as they saw them in their own countries, so as to obtain a picture of the socio-economic effects and constraints of tropical forest management.

THE COMPONENTS OF HUMAN DEMAND

Human demands on the forest fall into two categories. First, a demand for forest quantities; for the actual things a forest can produce: for timber, for food and for space for cultivation or for grazing; in all these man functions as one of the decomposer species. Secondly, a demand for forest qualities; for the effects that forests have on the environment of man—to protect supplies of water, provide havens for wildlife and maintain the pool of genetic resources, to protect the soil against erosion, and to provide space for recreation. In some of these man can act in symbiosis almost as part of the complex of synthesizer species, as for example when, by adding nutrients to the soil, he causes an enhancement of photosynthate production, either directly in the plants benefiting from the addition, or indirectly through increased nitrogen-fixation by legumes and other nitrogen-fixers. That the different demands may often be in conflict is illustrated in the subsequent pages. In almost every country in the world governments have attempted to control these demands and to regulate the interaction between them by legislation. In Chapter 18 of this book an attempt is made to consider the desiderata of such legislation.

The demand for forest products also has two components—a national demand and a local demand—and these may not necessarily be in harmony. The national demand is for timber, for defence, for housing, for construction, and increasingly as a raw material for industry, as for papermaking. Governments have looked to ways to ensure sufficient and continuing production of those forest products they needed; either for use or for export. Thus the earliest plantations of forest trees in India were of teak for the British Navy. In Indonesia timber ranks second as export income-earner only to oil (Chapter 8). In India, as Gupta records here (Chapter 7) the classification of some forest areas as 'revenue' forests by the state authorities has resulted in their almost total disappearance as forests, as they were cut as a source of revenue. This aspect of forest use, the exploitation of the forest resource to pay for other developments, is commonplace in the developing countries of the tropics. In this, tropical forest management is following the pattern of exploitation that existed in the forests of more temperate lands in an earlier period. The forests of the Weald in southern Britain were cut down to provide charcoal as fuel for the iron-smelters of the pre-Napoleonic wars, and in western Australia the forests were cleared of their giant jarrah trees to build bridges and provide sleepers for the expanding railway systems of the Victorian era. Whilst there is no doubt that such a system of cut-and-use does provide a boost to the local economy and provide employment for some of the local people, it can also have deleterious effects on their livelihood (see Chapter 4) and by destroying the habitat for wildlife, can cause damage to tribal and religious customs (see Chapter 15).

Local demands on the forest are for fuel-wood, food production and medicinal products, as well as for products associated with tribal or religious ceremonies.

Wood is the principal, and sometimes the only, source of fuel for perhaps one-quarter of the world's population. With the growth of awareness of the national need for the products of the tropical forests, governments attempted to get agreement with the inhabitants of the forests on the manner in which local demands could be met. In India, which has the best records of what was done, it seems likely that at the time the original agreements were made, the demands for fuelwood by the local inhabitants were not such as to cause concern to the foresters. With the enormous increase in population in the rural areas of the tropics in the last 50 years, and the reduction in the total forest area, however, the situation has changed. It has been estimated that the officially approved figure for wood to be cut for fuel would supply only about one-ninth of the needs of people. The balance presumably comes from illegal gathering of firewood. The problems highlighted by the people's agitation for access to neighbouring forests, and the government's response to this are described for one area in India by Gupta (Chapter 7).

Attempts are now being made in different parts of the world to meet the need for firewood, and the triple lines of Chinese poplars lining the roads in China, or the strip plantations of eucalypts that line every road leading out of Delhi, are encouraging indications of what can be done. In India there has been a very effective introduction of tree culture—it can hardly be described as forestry—into the irrigated lands of Haryana, where $3\frac{1}{2}$ million seedlings of a eucalyptus hybrid were distributed free to small farmers in 1980, and 5 million were to be available in 1981. These seedlings, planted along the banks between the irrigation bays, give millable timber in 8 years (Hooda, 1981). As a result they have been sold for timber, and so to date have contributed little to the fuel supply, although contributing greatly to the economy of the small farmers concerned (Hooda, 1981).

The forest contributes to the food needs of the forest-dwellers directly in terms of honey, fungi, fruit and nuts—among the non-timber products of forests. In some cases the fungi may make a disproportionate contribution to the trace nutrient intake of the people, and the lack of such components from the diet, owing to the reduction in facilities for collection following complete or partial exclusion of the tribal people from the forest, may have consequences out of proportion to the proportion of food contributed. Apart from these 'collected' foods, the so-called practice of 'shifting cultivation' has been endemic in every tropical forested area in the world except perhaps those of Amazonia. This practice, known also as 'swidden', 'jhuming' or 'cut-and-burn', consists of clearing and burning small patches of the forest and growing crops on then for 2 or 3 years and then allowing the land to revert to forest again. Subsequently it will be cut and burned again and the cycle repeated. Provided the time interval between successive burns is long enough, the plots are small enough, and sufficient large trees are left standing to provide some protection against rain damage and also seeds for new growth, the system is probably the best that can be adopted to prevent erosion on sloping land (Lal, 1980, personal communication). As the demands of an increasing population make the cycle shorter and shorter, damaging erosion and loss of topsoil takes place, and yield per unit area falls. There are wide variations in practice, even in limited areas (see Chapter 6).

Various schemes are being tried in different parts of the world to introduce a more stable agricultural or forestry component into the lives of the people, and several such are described in the following pages. In addition there is now a deliberate attempt to introduce the concept of agroforestry into those parts of the world where it had either never been practised, or where the technique had been forgotton. Some of the possibilities and prospects for the system are discussed later, (Chapters 8–11).

The third demand on the forests of the people living in or on the fringes of the forests is, in certain parts, for grazing for their livestock. Thus in the Ambala forest west of Delhi, India, some thousand Gurjars have the right to graze their buffaloes, which supply milk to local cities. Once again, the pressure of grazing was not so serious when the right to graze was originally accepted. The impact of increasing numbers of livestock is discussed later by Gupta (Chapter 7). As this is in a monsoon area the forest grasses dry up in the dry period, and during that time the herdsmen feed their buffaloes on branches lopped from the standing timber (Hooda, 1981). In theory, the forest-dwellers have the right to lop two-thirds of the way up the tree, but this right is almost impossible to police, and often lopping goes right to the top, with a consequent sharp reduction in timber production.

So much for the quantitative human demands on forest resources. All of them represent a decomposition of part of the biomass, and in some it occurs in a manner that will reduce the ability of the forest to regenerate and so maintain biomass production.

The human demands on forest qualities are easy to state, but difficult to quantify. Broadly speaking there are demands to maintain forest as forests for several reasons:

(1) as a protection of a watershed against erosion, because otherwise the transported eroded products may cause damage downstream, especially by silting water courses or reservoirs;
(2) as a means of maintaining the quality and quantity of water stored in a catchment;
(3) as a reserve for wildlife;
(4) as a biosphere reserve, necessary to preserve undisturbed, a part of the world for the importance of the flora and fauna contained in it as part of the pool of the world's genetic resources;
(5) lastly because forests have a recreational value, as places where the citizens of urban areas can come for rest, relaxation and refreshment—this is the forest quality most difficult to quantify, for what pleases one group of people may be anathema to another, and the decisions are made more difficult because people's views as to what constitutes a desirable environment change with time.

The extent to which any of these are taken into account depends, in a democratic community, on the impact they have on the people. Change in either quantity or quality of a water supply usually produces an immediate reaction from the people who use it anywhere in the world. The loss of a wildlife habitat is, in the developing countries, probably only felt by the forest-dwellers concerned and by a small proportion of the urban population. The loss of part of the world's gene pool is likely to worry only a small part of the countries' scientific community, and the readiness with which areas have been proclaimed as biosphere reserves in the tropics is perhaps surprising. The effect on the lives of the human inhabitants when an area is designated as a biosphere reserve does not yet appear to have been the subject of investigation. One can consequently only speculate. Certainly wildlife conservation schemes, and the creation of national parks, can lead to conflict with the inhabitants of the area unless their views and needs are taken into account in formulating the policies under which the parks are controlled (Lusigi, 1981).

The use of forest areas for recreation for the urban population also poses problems of conflict, and although it seems probable that the problems that have arisen in the developed countries of the temperate zone have not yet occurred in tropical forests, they seem likely to arise soon in those parts where development is proceeding fastest, as in the Philippines and India. If it is possible to generalize on a world scale it seems that for recreation people prefer an open forest with gaps between the trees where the plants and flowers of the undergrowth can flourish and be appreciated by the visitors. The dark and almost plantless floor of a closed forest, particularly of the single-species forest characteristic of plantation forestry, is generally disliked for recreation. Such forests, particularly where exotic species have been used, have been the subject of vigorous criticism. It was interesting to hear in India the same sort of criticism of the use of 'exotic' species such as eucalypts, as was heard of the use of *Pinus radiata* in the original home of eucalypts, Australia. The critics in both countries, and elsewhere, ignore what is common to all elementary courses on ecology—that animals are one of the major methods of seed distribution, and hence of species spread. Man merely contributes to spread of the species more effectively than other animals, and certainly more deliberately. He may also, by the production of fertile hybrids, be deliberately contributing to speciation.

There are many problems for forest management arising from the relatively unrestricted access of people into forests, and these often become more serious when the people are urban-dwellers, who may not appreciate the effects their actions can have on the forest. In the forests of the monsoon areas the risk of forest fire is always present in the dry season, and the control of access to restricted viewing or camping areas is often difficult to maintain. The entry of new diseases or pests into forests, associated with human factors, is another aspect of the effect of Man in causing change in the biotic component of the system, albeit on occasion accidentally. Fortunately at present there seem to have been no examples of such an effect in a tropical forest. The devastation in the jarrah forests of warm temperate western Australia, however, that has followed the introduction of the fungal pathogen *Phytophthera cinnamomi* serves as a warning of what could happen, for the distribution of that disease, now clearly visible on aerial photographs, follows closely the roads and tracts that modern settlement has cut through the forest.

The pressure of demand on the existing forests can be reduced by the development of alternative methods for growing trees, as for example by making plantations of single species in areas of presently damaged or eroded lands, or the development of 'community forests' under the control of the local people, and operated for their benefit, that is being assisted by the operations of FAO (see Chapter 18). The deliberate planting of trees as part of the agrigultural system—now becoming known as agroforestry—is perhaps an example of 'forest management' (see Chapters 9 and 10). There are still very few sites around the world where improved forest practices—in terms of timber yield per hectare—are being reconciled with the needs of the dwellers of the forest fringes, or with the environmental needs for water, wildlife and recreation.

A message comes through in the papers collected here, and expressed succinctly by the Inspector-General of India's forest, that there is need for replanning of the training of foresters to enable them to take the socio-economic aspects of forestry into account. There is perhaps an even greater need to develop a system whereby the urban masses, whose votes will control future development, can be educated to understand the nature of the problem, and the impact of their decisions, for as Ghosh points out (Chapter 3):

> It is patently clear that the future course of resource development and environmental management will be determined by urban-oriented citizens whose opinions prevail through their representatives who dominate the legislatures.

REFERENCES

Burges, A., and Drover, D. P. (1953). The rate of podzol development in the sands of the Woy Woy district, New South Wales *Aust. J, Botany*, **1**, 83–94.

Hissink, D. J. (1938). The reclamation of the Dutch saline soils and their further weathering under the humid climatic conditions of Holland, *Soil Sci.,* **48**, 83–94

Hooda, D. S. (1981). Field excursion of Workshop on *Socio-economic Effects and Constraints in Forest Management*. Haryana Forest Department, Ambala, India.

Jenny, K. (1941). *Factors of Soil Formation.* McGraw-Hill, New York.

Luisigi, W. J. (1981). New approaches to wildlife conservation in Kenya, *Ambio,* **X**, 87–92.

Malajczuk, N. and Bowen, G. D. (1974). Proteoid Roots are microbiologically induced. *Nature* **251**, 316–317.

Salisbury, E. J. (1925). Note on the edaphic succession on some dune siols with special reference to the time factor. *J. Ecol*., **13**, 322–8.

Thompson, C. H. (1981). Podzol chrono sequences on coastal dunes of eastern Australia. *Nature,* **291**, 59–61.

Socio-economic Effects and Constraints in Tropical Forest Management
Edited by E. G. Hallsworth

2

Some ideas for an interdisciplinary programme on world forests

ARTURO GÓMEZ-POMPA
National Institute for Research on Biotic Resources, Xalapa, Veracruz, Mexico

In ancient or modern primitive societies, man is a part of the natural forest ecosystem from which he obtains his food, medicine, construction materials and energy. With the advent of agriculture man found in the forests the fertile soil necessary for cultivating crops. His discovery that the soil has limited fertility and must be left fallow before the forest flourishes again restoring soil nutrients, led to agricultural nomadism. Primitive man greatly respected the forest to the point of endowing it with supernatural powers (the forest is still worshipped in some places).

Man's cultural evolution has led him to explore many roads. As a consequence his apparent bonds with the forest have been broken and lost to the point that he has assumed the role of an almighty being. Industrial societies have reversed the roles. Now the forest is just another component of the socio-economic system and man has become the dominator of ecosystems rather than being dominated by them. The industrial revolution supported by science and technology has made man the most important transforming agent in the biosphere.

Nevertheless, the restrictions imposed by the limited evolution of the human brain during the short time-span of the agricultural, industrial, scientific and technical revolutions have produced inconsistent situations which have not been sufficiently understood.

Industrial societies have developed regulatory political and economic systems which apparently ought to be responsible for counteracting man's impact on nature. However, until now no socio-economic system, be it socialism, communism or capitalism, has been able satisfactorily to resolve the contradictions between man's cultural evolution and the unfavourable environmental changes which in turn have repercussions on man himself. The same man capable of freeing the energy of the atom, poisons bodies of water and is responsible for the death of millions of men in fratricidal wars. In many regions of the globe this same man endowed with equal intellectual ability still lives as a hunter and harvester in the same primitive conditions as the ancestors of industrial society's man.

The cultural evolution of man could not have been successful without the massive transformation of natural ecosystems into different artificial ecosystems: the agricultural, cattle-raising, urban, etc., and the discovery of fossil energy sources of great magnitude—coal and petroleum.

The natural productivity of past and present natural ecosystems has been used to satisfy the human population's consumption needs. Today, expanses of land, formerly grazing lands or natural forests, are farmlands and summer pastures. Many rivers have been diverted and dammed to irrigate desert-like regions for agricultural production.

The process continues onward, and today we see another trend in many industrially underdeveloped countries: natural systems are transformed into artificial systems whose product yield is convertible into money which will theoretically permit entry into the industrially developed world. Unfortunately, human limitations have precluded a worldwide evaluation of this process and an attempt at understanding the ecological costs of this transformation of the earth's surface. Obviously we will not return to worshipping the natural forest. However, its massive destruction is a cause for concern due to our ignorance of the unforeseen consequences.

All of the foregoing points out the real need for a worldwide interdisciplinary study on forests and their transformation.

Without a doubt, one of the most important results of man's action on our planet is the modification of natural ecosystems and their transformation into diverse ecosystems: annual agricultural systems, perennial agricultural systems, grazing lands, urban areas, industrial areas and secondary ecosystems resulting from abandoned previously modified areas. Perhaps man's most important action on the biosphere on a worldwide scale can be strikingly observed in the transformation of forest ecosystems. At times the transformation has been qualitative and consequently less conspicuous, as in the case of forest plantations.

In human history different cultural attitudes towards forests have been recorded. Even in our era, different human groups understand forests in differing ways, attributable perhaps to cultural heritage or intercultural influence. Little attention has been paid to the understanding that different cultures have of the forest complex. An indigenous African group in the tsetse fly infestation area will have a very different perception of the importance of a jungle forest from the view held by a Finnish community devoted to the utilization of forest plantations. In the first case, the native community will see the forest as threatening tsetse fly control. The use of the wood for fuel and construction will be perhaps a secondary consideration. On the other hand, the Finnish community will see the forest exclusively as a source of income.

Focusing on the different societies' comprehension of forest ecosystems in time and space will surely shed light on the future of the world's forests. To illustrate this last point, we will mention the importance of forestry in Mayan culture. Forests were improved in a practical sense and converted into plantations containing a variety of trees for different uses (Barrera *et al.*, 1977). Another example of forest utilization linking culture and ecology is the nomadic agricultural system used on a large scale in various tropical countries. The farmer fells and burns his forest in order to grow crops on the land. After a certain number of years, which varies according to the place, planting stops and an alternative process is allowed to take place. Thereby, a secondary forest is established whose function is the restoration of soil nutrients through cultivation of different species of trees. After an approximate 5- or 10-year period the farmer uses part of the secondary forest's products and again fells and burns, repeating the operation from generation to generation. This process clearly illustrates that native cultures using it have a fairly profound understanding of soil fertility dynamics and primary and secondary tree manipulation. England gives us still another example. A tree or thicket when protected by law cannot be cut down without a long legal process.

In general, trees have played a part in the evolution of societies. They have even come to symbolize processes of social evolution. For example, in Maoist China reforestation was identified with a political movement to repair damage caused by the indiscriminate deforestation of former Chinese regimes. In this sense each commune's reforestation programme is a reparatory action against 'damage' previously done. This type of national action has been observed relatively recently in various countries. We can mention another example in the strong pressure exerted on Brazil to prevent the destruction of the Amazon rainforest. Still another example is found in the United States' forest conservation programmes imitated in other countries in the form of conservation programmes through national parks.

With advances in science and technology, knowledge of the forest's importance to the biosphere improves. Given the rate of world forest destruction resulting from expanding agricultural frontiers, issues of worldwide importance to the future of our planet have begun to be raised. Perhaps the most important aspect recently under study is the role of forest ecosystems in the fixation of the carbon dioxide produced by fossil energy combustion. In the past, oceans were considered the storehouses of excess carbon dioxide. Recent studies suggest that this is not so (Woodwell *et al.*, 1978) and forests play a minor role in carbon dioxide absorption. When forest mantles and leaf blankets are burned or decay microbically, carbon dioxide is released in the atmosphere. Each year the total amount of this gas increases due to lack of absorption systems. It is thought that the carbon dioxide liberated by forest destruction is almost equal to the amount released by fossil fuels. Several scientific articles have expressed different opinions on the subject, indicating that scientists are increasingly concerned about this process and the serious consequences it could have on climatic changes on Earth at a given moment. Obviously, a change in climate would have serious effects on large human population groups.

Unquestionably a global research programme will have to focus on several research areas simultaneously, with the participation of different disciplines. A historical perspective of forest evolution on our planet, both from a biological and historical point of view, will be necessary. The different cultural roles played by forests will need to be understood, as well as the role of forests in present-day cultures.

Processes of deforestation and reforestation will need to be analysed, based on this historical and cultural view in different climatic regions, and different economic and political systems. Both the rate and the causes of deforestation will have to be studied.

Deforestation processes have many distinct causes in Europe as compared with Brazil. Consequently, global solutions aimed at resolving the deforestation problem will need to be based on knowledge of these causes. In this case, several studies could be developed in different parts of the world in an attempt to understand the socio-economic, political and ecological processes involved in deforestation. In Mexico the process of felling trees in the wet tropics corresponds to a trend towards opening-up of new land for livestock. On the other hand, deforestation at high altitudes is brought about by landless peasants who have been relegated to the highest and steepest sections of the mountains, and whose only forms of sustenance are these uncultivated areas of sparse production. In these two extremes there is no question but that the only link between them is the action of deforestation. Nevertheless, there is likely to be an economic process behind it both in Mexico and other backward countries. Lack of government interest in protecting forests increases the problem. Perhaps the most important cause of forest destruction is the monetization of resources. To give a money value to a forest puts it on a par with other products, although there is actually no real equivalency.

As for reforestation, it would also be advisable to select several regions and to try to understand institutional or national motivation favouring reforestation operations.

It would also be important to develop parallel research projects on the possible ecological effects of the simplification of forest ecosystems through forest plantations. Without question, forest plantations have simplified forest biotic composition in order to adapt to available industrial processes. Nevertheless, possible secondary effects of biotic degradation of forest ecosystems on a global scale have not been evaluated.

The strategy to be followed in this field would involve the construction of a matrix which would include, on the one hand, the continents, climatic zones and types of vegetation in the world, and on the other hand, different kinds and levels of research of a biological, ecological, historical, economic, conservational and recreational nature. In this sense it would be necessary to evaluate as thoroughly as possible the role of forests as cultural and recreational resources. Apparently there are cultural and geographical differences with respect to the value attributed to forest ecosystems.

Another parallel study could be carried out on a global basis on the role played by forests as germinal plasma banks of plant and animal species. In addition, research on the real role of forests and their ecosystems in the biosphere's nutrient estimate would also be included.

It has also been thought that forest ecosystems could play a significant role as an energy source for future generations. Even though many societies have been utilizing bioenergy as a basic source, this resource has hardly been studied with a wider view towards the future. A comparative study of the short- and long-term real energy potential of different kinds of forests by geographical regions could be of great consequence and would add to the value of a global study of these problems.

In an international project like the one being proposed, involvement of organizations having international coordination functions with widespread participation like MAB–UNESCO, IUFRO and SCOPE, would be desirable. The possible independent participation of a select group of prestigious universities and institutes in an interdisciplinary programme on world forests would be extremely valuable, since intergovernmental organizations are often considerably restricted. On the other hand, human and economic resources are not always available simultaneously.

REFERENCES

Barrera, A., Gómez-Pompa, A. and Vazquez Yanes, C. (1977). 'El manejo de las selvas por los Mayas,' *Biotica,* **2**, 47–60.

Rappaport, R. A. (1972). 'Forests and the purposes of man'. Proceedings Symposium on Fire in the Environment, *US Forest Service Publication 276*, pp. 37–46.

Woodwell, G. M., Whittaker, R. H., Reiners, W. A., Likens, G. E., Delwiche, C. C. and Botkin, D. B. (1978). 'The biota and the world carbon budget,' *Science*, **199**, 141–146.

Socio-economic Effects and Constraints in Tropical Forest Management
Edited by E. G. Hallsworth

3

Socio-economic effects and constraints in forest management

R. C. GHOSH*
Director, Forestry Research, Forest Research Institute

India, the seventh largest country of the world, is a distinct geographical unit situated at the head of the Indian ocean. Lying between the latitudes 8°N and 38°N (the Tropic of Cancer runs almost halfway) and longitudes 68°E and 98°E, it covers an area of over 3.2 million km^2. About 23 per cent of the land surface is occupied by forest and 50 per cent by agriculture, and the country has a human population of about 625 million. The country is separated from the rest of Asia on the north by the Himalayas. Broadly, it consists of a peninsula to the south of the Tropic of Cancer and broad alluvial plains and the high mountains to the north. The peninsula comprises a plateau of comparatively low elevation flanked by fairly high hills of the Western Ghats from which the land slopes eastwards. The Eastern Ghats form the eastern boundary of the plateau. To the north of the peninsula is the low plateau of central India gradually sloping down to the broad lowland valleys of the Indus, the Ganges and the Brahmaputra. To the north of this extensive plain is the high barrier of the Himalayan mountains.

The Indian subcontinent presents a wide range of ecological situations, climatic conditions ranging from the hottest to the coldest and from the driest to the wettest. The distribution of rainfall is largely governed by the relief of the land and the Himalayas in the north play a vital role in this respect.

The terrestrial vegetation has evolved in equilibrium with the changing contours of the coastlines, the upheaval of land masses, the formation and dislocation of land bridges, the direction of ocean currents and wind leading

* Dr Ghosh is now at the Office of the Chief Conservator of Forests, West Bengal, P-16 India Exchange Place Extension, Calcutta.

to migration, isolation and evolution culminating in the typical floral diversity and distribution in the subcontinent.

During the past centuries and the early decades of the current one, the forests owned by the states and the private individuals came under the pressure of extensive agro-economic movement and suffered widespread trespass and disintegration as a result of which, except in the remote and inaccessible parts of the country, original form and structure of the forests have been modified by the successive events of human history. Land-use patterns in the subcontinent have developed on the basis of physical and climatic conditions as well as being influenced by growth and migration of population in the different periods of history. Deforestation in the Indo-Gangetic plain was carried out by the Aryans since the time of their migration that took place between 2000 and 1800 B.C., approximately. They were partly agricultural and partly pastoral people, and they required land for cultivation and pasture apart from their needs for construction for numerous settlements. The main impact has been felt through a number of routes such as clearing, burning, grazing, abandonment due to loss of soil or its fertility, lopping of trees to extinction, etc. The vegetation and the environment has changed, sometimes to human and environmental gains, occasionally to the benefit of one of them, and more often to the detriment of both.

Demarcation of forest areas was followed by certain legal injunctions on reckless clearing of Reserve Forests. Nevertheless, extensive marginal forests were left out adjacent to the village sites, and for many of the social needs these forests used to be the chief source. As time passed, agriculture advanced at the expense of forests and the residual vegetation began to degenerate due to intense exploitation. Even when reduced to scrubs, these forests provided the major requirement of fuel and other commodities to the millions in the countryside.

The forests as they survive today have also a very uneven and unbalanced distribution territorially, and this was essentially influenced by the extent and direction of human settlement and agricultural expansion. Large areas in the Indo-Gangetic plain, in the coastal strips and other fertile systems are either devoid of forests or carry only sparsely distributed trees and degraded vegetation. Economic and environmental contribution of forestry varies widely from region to region depending on the market demand, utility of different species, their value, regional socio-economic conditions and the geographic situation in which they occur.

Forest areas under different forest types are shown in Table 2. For obvious reasons, no claim is made that the figures of forest area reported under different forest types in this chapter are very accurate, but they are probably the best that could be obtained with the available data and are fairly reliable for any broad-based planning.

Considering various economic environments in which villages are situated,

Table 1 The forest areas of India in relation to population

States/union territories	Area (thousand ha) Total area	Forest	Forest area as percentage of total	Forest area (ha) per capita
Andhra Pradesh	27,682	6,480	23.41	0.14
Assam	7,852	2,855	36.36	0.23
Bihar	17,388	2,932	16.86	0.05
Gujarat	19,598	1,697	8.66	0.06
Haryana	4,422	152	3.43	0.01
Himachal Pradesh	5,567	2,167	38.93	0.61
Jammu & Kashmir	22,224	2,072	9.32	0.43
Karnataka	19,177	3,608	18.82	0.12
Kerala	3,887	1,127	28.99	0.05
Madhya Pradesh	44,284	16,835	38.01	0.38
Maharashtra	30,776	6,613	21.49	0.12
Manipur	2,236	602	26.92	0.53
Meghalaya	2,249	699	31.08	0.66
Nagaland	1,653	288	17.42	0.54
Orissa	15,578	6,793	43.61	0.30
Punjab	5,036	213	4.23	0.01
Rajasthan	34,221	3,589	10.49	0.13
Sikkim	730	265	36.30	0.73
Tamil Nadu	13,007	2,234	17.17	0.05
Tripura	1,048	605	57.73	0.37
Uttar Pradesh	29,441	5,013	17.03	0.05
West Bengal	8,785	1,183	13.47	0.03
Union territories				
Andamans & Nicobar Islands	829	746	89.99	6.16
Arunachal Pradesh	8,358	5,154	61.67	10.52
Dadra & Nagar Haveli	49	20	40.82	0.26
Delhi	149	4	2.68	Negligible
Goa, Daman & Diu	381	131	34.38	0.14
Mizoram	2,109	887	42.06	—
Others	63	—	—	—
All India	328,779	74,964 or 75,000	22.80	0.13

the following categories can be broadly identified:

(1) villages mainly dependent on forests and marginally on agriculture—e.g. in the Himalayas and similar production zones;
(2) villages totally dependent on agriculture, as in the fertile river basins;
(3) villages subsisting on agriculture and pasture economy, as in the dry and semi-arid tracts.

Table 2 Areas of the various forest types of India

Type-group no.	Forest type	Area (million ha)	Percentage of total forest area
1	Tropical wet evergreen forests	4.503	6.0
2	Tropical semi-evergreen forests	1.854	2.5
3	Tropical moist deciduous forests	23.303	30.9
4	Littoral and swamp forests	0.671	0.9
5	Tropical dry deciduous forests	29.154	38.7
6	Tropical thorn forests	5.236	6.9
7	Tropical dry evergreen forests	0.075	0.1
8	Subtropical broadleaved forests	0.287	0.4
9	Subtropical pine forests	3.740	5.0
10	Subtropical dry evergreen forests	0.173	0.2
11	Montane wet temperate forests	1.613	2.1
12	Himalayan moist temperate forests	2.725	3.6
13	Himalayan dry temperate forests	0.227	0.3
14	Sub-alpine forests }		
15	Moist alpine scrub }	1.790	2.4
16	Dry alpine scrub }		
	TOTAL	75.351 approx.	100.0

Thus it is obvious that there are contrasting situations; thickly populated areas of the river basins where forest lands have been diverted for production of food, and on the other hand there are regions of extensive biological capital with low population density. Between these, there are many peculiar intermediate situations.

In the north-eastern region, shifting cultivation is an age-old practice which is rooted in the cultural and religious ethos of the tribal communities. A low investment system of subsistence agriculture, it has gone out of step due to constraints of the environment and population growth which have reduced the cycle of cultivation and fallowing. Now, there is not enough area for the system to function properly, and degradation is rampant. Various estimates exist of the areas under shifting cultivation and the population involved. The total area involved is around 0.5 million ha supporting a population of about 0.3 million tribals.

Similar to nomadic agriculture is nomadic pastoralism which is practised in many parts of India. Today it is only a remnant of the pre-agricultural economy which has been relegated to the lowest strata of the social and economic complex. A typical area is the district of Mutch in Gujerat, and the areas around have the same history of how forests have suffered widespread destruction. The residual forests are either of grasses or thorny scrubs and small trees of *Acacia* spp., *Prosepis* spp., *Salvadora* spp., *Tamarix* spp., etc. It is one of the critical areas where the problems are extremely complex. Saline and alkaline soils cover a large area in this tract. If one takes a look at the rural scene, the economy of the region is dominated by two major uses of land—agriculture and pasture. Time and again this region has been overtaken by calamities following periodic droughts and the economy has warped. Destruction of forests by reckless clearance for agriculture and excessive use for pastoral purposes have brought extensive ruination.

Studies were conducted by Indian Space Research Organization, Ahmedabad, in which 222 villages were surveyed to find out the losses on account of droughts which are frequent in this tract. Tables 3 and 4 clearly show how animal husbandry suffered from periodic droughts and how water famine led to a large reduction in animal population. Unfortunately the same activities, i.e. agriculture and grazing, are still considered as props of economic growth and development despite the fact that both these professions have been fatal liabilities in the past.

The arid area in India covers 208,487 km^2 (61 per cent of the Rajasthan state) with a population of 9,803,000. This tropical desert is characterized by scarcity of water, high evapotranspiration, low and uncertain precipitation and an inhospitable sand-dune terrain. Evidently the arid region is unattractive to human occupation though the inter-basin transfer of water from the Himalayan rivers to the desert sands has modified the environment in certain tracts of the state, making land productive for agriculture. But this is a part of the story. The age-old agropastoral economies are fast decaying, yielding place to subsistence existence in which cultivation of food crops dominates the land-use pattern. There is virtually no socio-economic cohesion between the two juxtaposed communities, the pastoral nomads and the settled farmers, and often they remain in conflict. Even the contact within the arid land among the pastoral folks, cultivators and urban-dwellers is marginal, which perpetuates disintegration of the society and the land-use systems. In order to reduce the inherent risk of droughts, silvipastoral development and livestock industry should constitute the basis of land-use and irrigated cultivation ought instead to take a supporting role to this task.

Table 3 Land utilization in several villages

	Land utilization (ha)			
Village	Cultivated	Grazing	Waste land	Forest
Narayan-Sarovar	3807.00	980.00	30338.00	
Baranda	3092.00	223.00	32912.00	2360.00
Siyot	7146.00	332.00	18992.00	113.00
Dayapar	11369.00	1185.00	39961.00	5637.00
Naliya	7776.00	2742.00	2389.00	243.00

Table 4 Animal population in villages shown in Table 3

Village	Animal population									
	Cows		Bullocks		Buffaloes		Goats		Sheep	
	Past	Present	Past	Present	Past	Present	Past	Present	Past	Present
Narayan-Sarovar	4,236	1,951	760	263	725	165	8,107	3,980	1,157	1,172
Baranda	5,180	1,231	83	116	4,160	272	3,600	6,310	560	885
Siyot	6,800	1,935	280	272	4,060	593	3,200	5,730	1,615	275
Dayapar	23,566	2,989	982	567	2,819	846	13,250	7,724	2,600	1,400
Naliya	2,870	607	918	331	1,058	268	3,400	3,102	800	956

The National Commission on Agriculture has identified about 30 million ha of hardwood forests and about 1 million ha of conifer forests available for conversion to highly productive plantations which would guarantee supply to various industrial units. In the northern Himalayan zone the forests start in the foothills, sometimes as pure crops and often as stands constituting a complex broadleaved association of several species. Higher up, pines dominate, gradually yielding place to fir and spruce. From the utilitarian point of view, the number of important species are spruce (*Picea smithiana*), fir (*Abies pindrow*), deodar (*Cedrus deodara*) and pines (*Pinus roxburghii* and *P. griffithii*) among the conifers, and walnut (*Juglana regia*), birch (*Betula alnoides* and *B. cylindrostachys*), ash (*Fraxinus excelsior*), elm (*Ulmus wallichiana*), maples (*Acer* spp.), alders (*Alnus* spp.), poplars (*Populus* spp.), hornbeam (*Carpinus vimenea*), and sal (*Shorea robusta*) among the broadleaved.

Down south in the peninsular area, teak is the most important species and also rosewood (*Dalbergia latifolia*). In the wet evergreen forests, Dipterocarps constitute the most important crop. The species that have made a mark in the field of veneer industry are *Tectona grandis, Michelia champaca, Schima wallichii,* Dipterocarps and Terminalias. Future industrial possibilities depend on the large-scale plantations of these species and there is no alternative site except what has remained within the boundary of the old reserves situated in these regions. In fact there is hardly any option, and it is difficult to manipulate management practices to cater, in addition, for the requirements of millions of rural population without upsetting the principal functions of such forest areas.

Grazing incidence in forests for the whole country is indicated in Table 5, which shows that since 1956 there has been a marked increase in the number of animals, except camels, being grazed in the forests. These figures, taken in conjunction with the increasing demand for wood and pulp products (Table 6) and the continuing demand for fuelwood (Table 7) emphasize the dilemma facing Indian forest management.

The objectives of management of forests for production purposes have to be clearly understood with due emphasis on ultimate utilization of raw material and, therefore, on specific requirements of industries concerned and the markets to be sustained. In the past, management was primarily linked with limited local demand, and in the absence of industrial growth large users did not exist. Forestry has moved away from this situation and investment opportunities in the potentially productive areas have come to be recognized and the rationale is binding on the foresters to intensify efforts to stimulate higher production on as many sites as possible. It is necessary to change over to industrial species, especially for production of pulp and panel products, and when sufficiently large areas are converted into commercial crops, only then can the anticipated demand be met and economic growth sustained.

Historically, in India, forest management aimed at protection and conser-

Table 5 Grazing stock in the forests ('000)

Animals (1)	Past position[1]			Current position		
	Total population as per 1956 census (2)	Animals grazed in forests 1957–58 (3)	Percentage of col. (3) to col. (2) (4)	Total population as per 1972[2] census (provisional) (5)	Animals grazed in forests 1973–74[3] (6)	Percentage of col. (6) to col. (5) (7)
Cattle	158,651	17,133	10.8	78,865	26,115	14.8
Buffaloes	44,916	4,303	9.6	57,941	9,892	17.1
Sheep	39,246 }	12,579	13.3	40,395	11,208	27.8
Goats	55,405 }			68,024	4,740	7.9
Camels	776	146	18.9	1,126	121	10.8
Other animals	2,642*	673	25.4	8,631	2,291	26.7

[1]1961. *One Hundred Years of Indian Forestry,* vol. II, p. 57 (Table 16), Dehra Dun, FRI.
[2]Eleventh All-India Livestock Census, 1972 (provisional figures). New Delhi, Directorate of Economics and Statistics, Ministry of Agriculture and Irrigation, Government of India.
[3]Chief Conservator of Forests. The figures exclude those for Karnataka, Kerala, Manipur, Meghalaya and West Bengal, from where figures were not available.
*Excluded pigs.
Note: separate data for sheep and goats for 1957–58 not available.

Table 6 Aggregate raw material requirement for A.D. 1980, 1985 and 2000 under assumption of high income growth (in '000 m^3)

Item	Coniferous wood			Hardwood			Total wood		
	1980	1985	2000	1980	1985	2000	1980	1985	2000
Sawnwood	2,020	2,600	4,020	12,080	15,700	25,630	14,100	18,300	29,650
Panel products:									
Plywood and veneer	120	190	430	480	775	1,725	600	965	2,155
Fibreboard	20	25	40	75	100	160	95	125	200
Pulp and paper	1,100	1,615	5,975	3,075	4,440	11,720	4,175	6,055	17,695
Matchwood	—	—	—	535	680	1,415	535	680	1,414
Roundwood	1,480	1,810	2,665	5,910	7,245	10,670	7,390	9,055	13,335
TOTAL	4,740	6,240	13,130	22,155	28,840	51,320	26,895	35,180	64,450

Table 7 Recorded outturn of forest products (1972–73 to 1976–77)

Sl. No.	Category	1972–73	1973–74	1974–75	1975–76	1976–77	1977–78*
1.	Industrial wood:						
	(a) Saw logs, veneer logs and logs for sleepers	4,503 (17.4)	5,068 (19.3)	5,132 (19.35)	5,235 (20.22)	5,614 (21.2)	
	(b) Other industrial wood not exactly specified	5,180 (20.1)	4,640 (17.7)	4,666 (17.59)	5,537 (21.38)	5,494 (20.75)	
	TOTAL	9,613 (37.5)	9,708 (37.0)	9,798 (36.94)	10,772 (41.6)	11,108 (41.95)	
2.	Fuelwood (including wood for charcoal)	16,143 (62.5)	16,534 (63.0)	16,722 (63.05)	15,122 (58.4)	15,372 (58.05)	
	TOTAL WOOD	25,826	26,242	26,520	25,894	26,480	
	Of which;						
	Coniferous wood	1,670 (6.5)	1,669 (6.4)	1,339 (5.05)	1,543 (5.96)	1,772 (6.69)	
	Non-coniferous wood	24,156 (93.5)	24,573 (93.6)	25,181 (94.95)	24,351 (94.04)	24,708 (93.31)	

*Statistics not yet collected. Figures within parentheses indicate percentages.
Source: Central Forestry Commission.

vation. In fact the foresters were the first to plead for environmental conservation. Many of the forests in the country are situated in the catchments of many important rivers, which are highly sensitive to any kind of interference; development of irrigation, power-generation and flood-protection depend on the stability of these areas where, though the productivity is high, there is a risk of other utilities being sacrificed.

A reinterpretation of the goals and status of forestry in the framework of land-utilization is necessary to replace the concept of an autonomous role of forestry based on the production function alone. The new implications have have to be studied in great detail to judge the propriety of the management alternatives and finally choose the one that would cater for, apart from wood, many other amenities that are bountiful in an ecologically stable environment.

This calls for caution, and to relieve the burden on such forests opportunities must be created for growing trees on new sites and in that respect 'social forestry', as conceived in India (which includes production of fuel and small timber), has extensive scope.

The range of opportunity of social forestry has to be considered in relation to the large expanse of agricultural area in the country and a situation that would not easily permit release of land for large-scale cultivation of trees. Nevertheless tree lands have to be created by planting singly, in groups, in strips or small blocks without interfering with agricultural production. Thus alone, the energy resources can be built up and other necessities taken care of. About 13 per cent of the land area (43.0 million ha) is, however, potentially productive wastelands, a substantial part of which can be reclaimed through raising of tree crops. Railways in India cover a length of 57,790 km and the road system extends over 1,200,000 km. The national highways traverse over 2900 km. Total length of canals, distributories and minor channels extends over 150,000 km. It is estimated that the area available for tree growth along roads, railways and canals is about 902,000 ha, which is quite a significant area. Plantations over 50,000 row km and 33,000 row km have already been raised in Punjab and Haryana.

Such expansion of social forestry would be a bold step forward, but there are some indications that even a sustained programme of this size would be inadequate to meet minimum social needs and avert fuelwood crisis in some regions. For example, a study made in Madhya Pradesh in 1977 of present and projected fuelwood requirements indicates that 26 of the 45 districts of the state are already in a deficit position, and if the present fuelwood use and population trends continue, 39 districts would be in deficit by the end of the century. The Forest Department estimates that in order to meet the projected deficit and avert a serious energy crisis, a total area of 19,630 km^2 has to be brought under fuelwood plantations under social forestry programme by the year 2000, at the rate of 854 km^2 per year. The annual cost is estimated to be

Rs. 128 million and employment would be generated for about 60,000 people. Creation of employment opportunities for the large rural labour force is oe of the main planks of such a programme.

Against this background a decision-making mechanism has to be evolved for the future forest-management and land-use policy of the country. While production of food and other commodities should be an essential feature of such a policy, the fact that a conducive environment is equally important to sustain life should not be overlooked. As the country's economy is based primarily on agriculture, categorization of land is done only on the suitability for agriculture. But the continuous disturbance of the environment and physical alterations of the land surface create a situation where one cannot safely speculate on the economics of such land-use in the climatic conditions that prevail in this part of the world. At the same time social and political constraints do not permit a drastic departure from the current pattern and these are the main obstables to a high quality land-use programme.

The present era could be hopefully named the 'ecological era'. Land resource is the most essential part of the environment and its management should be based on the acceptance of the ecological characteristics which pervade the environment. At the same time, we have to be cautious that it does not degenerate into a bandwagon phenomenon and be misused while taking steps to prevent ecological ravages. Land has to be managed in harmony with all ecological principles and it is necessary to distribute the costs and benefits equitably on all generations, present and future, so that the nation gains a continuous fruitful life in which aspirations are fulfilled to the best possible extent. It is quite understandable that farmers and foresters alike are mainly interested to know the environmental conditions conducive to growing crops or trees. In this age when the goal is higher production no area is exclused and no profession exempt. It has already been stated that an extensive area of land has been reduced to poor quality due to continuous ill-treatment, and also due to lack of clear management objectives. Hence there is a growing recognition that long-term solutions must be sought. Future planned land-use, engineered by ecologically minded foresters, agriculturists and architects, respecting better quality and a standard of life based on aesthetics and ethics and devoted to improvement of our environment, gives us a new opportunity.

However, modern land-use should be designed to deal with many apparently conflicting uses. A new meaning has to be found for intensive land management and a purposeful policy adopted to widen the concept and scope of intensification which should take into account the other bounties that flow from a particular management practice. Generally, while taking management decisions, production occupies the central position, but it would not be inconsistent with some management practices that the central place is occupied by non-economic motives. In formulating land-use policy the needs of the society, as well as of the individual, must be met. The project must reflect the current state of social altitudes and it should be continuously monitored to eliminate irrelevant operations and to make room for new development.

If the land is to be used efficiently on a permanent basis, the distribution of uses must correspond with:

(1) its inherent capabilities, as determined through the appraisal of soil, topography and climate;
(2) its possibilities of improvement through restoration, conservation, irrigation, etc.;
(3) other factors influencing the land-use pattern, such as population pressure on land resources, population relocation, land tenure, watershed protection, road infrastructure, essential demands, distance to market, etc.

'The technical ideal would be to organize production in such a way as to obtain from each site class the most valuable product, without destroying or depleting the soil resources and without creating other environmental hazards in the area of influence (silting, floods, droughts, destruction of wildlife, etc.). Once the land-use pattern with its basic technical and sociological considerations has been agreed upon, it should be adhered to and not be altered at the whims of politicians or other people with local influence.' (FAO, 1978)

However, if the forestry programmes are to be efficiently implemented it will be necessary to cover many other aspects including extension work in the area under a project. These aspects are:

(1) estimating local needs for wood and fuel;
(2) identifying other forest products such as food, gums, tannins, etc.;
(3) income generation and industrial possibilities; and
(4) environmental aspects of forestry.

With increased understanding, wiser decisions can be made about the manipulation and exploitation of various ecosystems. It may be that the most urgent task for the planner is to construct a model of rural environment to show the conflict between many interests and to give guidance in deciding about the changes in structure and policies which would provide proper environmental security. This is a case for operational research.

To quote Sri S. K. Seth:

'the major objectives of developmental planning for forestry would be:

(1) to link up forest resources with forest-based industries on the basis of physiographic, industrial, and economic catchment concepts;
(2) to develop forestry as a support to rural economy and to trigger off development (through cottage, small, medium and large scale industries) in backward regions, where forests happen to be the most important local natural resource;

(3) to develop an infrastructure on the basis of developing an area, so that the costs of development are shared by the various sectors which will benefit from it—forestry is only one of such sectors, although locally, it may happen to be the most important one (Seth, 1978).

Such a sectoral plan would basically involve:

(a) the estimating of the future demand for forest products, with particular reference to the demand by the rural population;
(b) the assessing of the size and the nature of the existing land for the development of forest resources and its capacity to fulfil future needs;
(c) the .assessing of unutilized, underutilized and misutilized land which could be diverted to creating a forest resource (under organized 'forestry', social forestry, dispersed 'treeland', 'multi-purpose farmforests' and such other concepts);
(d) the determining of the nature and extent of a man-made forests programme to be undertaken and then finding valid socio-economic reasons to attract capital for implementing the programme;
(e) the identifying and the formulating of specific projects relating to forest-based industries, based on specific resource areas;
(f) the developing of the institutions which can play the desired role in modernizing the practice of forestry, including the generating and transferring of technology, the training of personnel, the creating of business, the managing of man, the managing of markets and promoting them.'

Governments must use their agencies and resources to direct the investment behaviour for basic management of forest lands towards social purposes. However, there are serious discrepancies between the conditions of management responsibility in the existing administrative structure and those actually prevailing on public lands. Forest Departments were developed as custodians and management agencies of more than 20 per cent of India's land surface on the broad assumption that these were under adequate forest cover, free from occupational rights and pressures to endanger regulatory controls. But the picture is different, and today, more than half the area is degraded, overgrazed and virtually under private control though the territorial jurisdiction continues to vest in the forest organizations. The rural scene, as we witness today, is the complex imprint of the past history. In any human society, development and growth is inherent and there is a constant endeavour to generate and enhance certain durable attributes of a viable social structure and a stable environment.

The forest environment as juxtaposed to society may be interpreted to include forest area, the rural surrounds and the interface between the two. In some respects these are three reasonably discreet areas, nevertheless interrelated and interdependent though often in conflict. In the process of growth and development in such a situation, ecological and productive dilemmas were often caught in the polarization of interests. What was supposed to be the centre of development, due to distortion of the process and social attitudes, became the heart of a perpetual conflict. The rural populace was divided into two major groups: the rural elite—the landholders—and the poorer people who were only partially assimilated into the agricultural life of the region. So far as forest produce is concerned there was the disrupting presence of the contractors' money-economy. If trees were to be preserved, contractors' interests should be the first to be safeguarded and this was very cleverly manipulated through the joint connivance of the urban trader and the affluent villagers who provided the middle man. The cumulative effect of these forces was the prosperity of the distant townships and only marginal benefits of the villages situated around the forests. Thus, while the growth was consumed by a section of the population, this created a surplus unemployed population which were thrown out of the zone of rural progress and development and dumped on the fringe of agriculture-dominated economy which incidentally is the area of forestry activities. Deprived of their share of development and growth, they could only look to forest-based economy for some subsistence existence. Time and again there has been an influx of such population into the forested areas. Now that saturation stage has been reached, the losers are those whose survival is at stake—the forests and the destitute population turned out of the main stream of village economy.

This is the predicament. What can a handful of forestry personnel do to oppose the action of thousands of people who constantly nibble at the forest resources? The most important technical lag is that of organizational techniques on which rests the protection and productive functions of the forests. We are expecting too much from too few who, though competent, are overburdened with their tasks. At the same time, much depends on good faith and goodwill of the people, and such goodwill can easily be dissipated if the parties had different perceptions on the nature of land development. People have genuine demands and their expectation from the use of a resource cannot be tossed away.

The need is to demonstrate what could be achieved if it is possible to combine social and productive objectives. Measures are required to ensure that all within the influence of a particular environmental situation are given sufficient stake in the environment to comply with the needs for protection and maintenance, and to forgo destructive forest uses through reckless exercise of their rights and privileges. Some replacement sites would be necessary to provide the needs now sought in the forests.

The concept of community forestry, or social forestry as it is called in India, is supposed to reduce destructive pressures on forest resources by providing economic alternatives, and unifies public and private interest in a compatible set of objectives. A successful operational programme would depend on investments both by government—in terms of technical assistance, planting stock and wage subsidies—and by villagers in form of forgone grazing and labour for protection and management. Agreements are required, and so also technologies to secure a sharing of benefits and obligations sufficiently to sustain the necessary investments over time.

At present few guidelines exist to suggest how the requirement of community forestry can be satisfied under different ecological, social and administrative circumstances. They may also include a scheme of insurance against natural hazards and price instabilities. The programmes would be directed towards reducing the costs and risks that are inherently greater in long-term than in short-term enterprises. What do these essential forms of participation imply for the administrative structure? Again, we can only consider possible criteria and examples rather than answers. Community forestry is undertaken in some states as an extra activity of territorial or developmental divisions in forest departments. This strategy assumes that the broadening of a department's professional capacities would increase overall effectiveness. This is too simple; in fact, what we need are:

(1) a structure which distributes skills and resources to their most efficient level of operation;
(2) a structure which provides authority, resources and incentives sufficient for satisfaction of objectives;
(3) a structure which feeds back experience into operations as quickly as possible.

Modification of the structure along these lines would result in increased efficiency of the overall operations. Unless the structure is such as can bend patterns of development towards social progress, justice and equity, any system would warp and finally collapse.

It is also necessary to examine the urban mind to analyse attitudes and to assess levels of knowledge regarding environmental conservation, development of natural resources and rural matters generally. It is patently clear that the future course of resource development and environment management will be determined by the urban citizens whose opinions prevail through their representatives who dominate the legislatures at the central and state levels. It is of paramount importance that the electorates and those they elect should be well-informed with respect to resource and environmental problems if the right policies are to be adopted and wrong decisions avoided.

REFERENCES

FAO (1978). *Forestry for Local Community Development.*
Seth, S. K. (1978). 'Forests and forestry in the Himalayan region.' *Proceedings, Seminar on the Utilization of Himalayan Resources.* Delhi.

Socio-economic Effects and Constraints in Tropical Forest Management
Edited by E. G. Hallsworth

4

Forest development and tribal welfare: analysis of some policy issues

RATNA MURDIA
Tata Institute of Social Sciences, Deonar, Bombay 400 088

INTRODUCTION

Tribals constitute 7 per cent of India's population. They are concentrated in the middle belt of India extending from Daman and Thana in Maharashtra on the western coast to west of Ganges in West Bengal passing through the border regions of Maharashtra, Madhya Pradesh, Bihar and Orissa. This belt joins the Himalayan tribal belt near Darjeeling, to the east of which lies the north-eastern tribal region covering all the states and union territories in that area. On the western end of the central belt the tribals are concentrated in hilly regions on the eastern border of Gujarat extended up to Udaipur in Rajasthan. The Dandakaranya region, the other major concentration in Middle India, lies to the south of the central belt and is separated from it by the plains of Chattisgarh. It extends from Phulbani in Orissa to Chandrapur and Yeotmal in Maharashtra in the west. Durg-Raipur region in Madhya Pradesh forms its northern boundary. In the south it spills beyond the Godavari valley into the plains of Khamman in Andhra Pradesh. Some tribal communities live in the Himalayan region, starting from Himachal in the west, through hills of the Uttar Pradesh forest of Terai and Sikkim joining the main belt in the hills of West Bengal and the north-east. The tribal population in the southern states is largely confined to the hilly tri-junction. The oceanic tribal groups are in the Andaman and Nicobar Islands in the Bay of Bengal and Lakshadweep in the Arabian Sea.

There are about 300 tribal communities in India which differ in size. On the one extreme are the Gonds who number about 4 million and on the other extreme are the Onges numbering only 112 and the Great Andamanese who are only 19. While some of the communities are homogenous, others like the Gonds comprise a number of smaller sub-groups some of which, for all practical purposes, may be distinct communities except for their generic name.

There are six states and union territories, viz.: Arunachal Pradesh, Meghalaya, Mizoram, Nagaland, Lakshadweep and Nagar Haveli which are predominantly tribal. In the bigger states like Madhya Pradesh, although the percentage of tribal population is only 25 or less, there are compact areas having a high degree of tribal concentration. The extent of concentration of tribal population varies from state to state, concentration being highest in Manipur and Nagaland and lowest in Karnatka and Goa.

The tribal communities are at different stages of economic development. On the one extreme are the communities like Jarawas and Sentenelese in Andaman and Nicobar Islands who live in isolation a primitive way of life. There are a number of small groups in several states, who are at the food-gathering and hunting stage with limited degrees of contact with the outside world. On the other extreme of the socio-economic spectrum are those tribal communities who live in the vicinity of industrial and mining complexes and are indistinguishable from other groups. The bulk of the tribal communities in India, however, are settled farmers. The next biggest group is that of shifting cultivators, who are widely spread in the north-eastern region and the hills of southern Orissa and northern Andhra Pradesh. There are small artisan groups, like the Asurs in Bihar or Kathodias in Rajasthan.

The tribal economy is largely undifferentiated. Each individual has a variety of basic skills which enable him to eke out a living from his natural habitat. He can identify the variety of plants and their uses; he is familiar with the behaviour pattern of wild animals and, in many areas, is a good trapper and marksman; he can build his own house with local materials; he is adept at using the axe and can shape the wood for a variety of uses; he can press oil seeds and preserve skins. His dependence on the outside world is limited and may be confined to purchasing salt, cloth and some odd articles. This situation, however, is changing. In many areas, tribals are now indistinguishable from other rural communities; the members of tribal communities are specializing as agriculturists and losing the versatile command over a wide spectrum of elementary skills.

The level of education among the tribals varies considerably from one group to another. The overall literacy level amongst the tribals is 11 per cent according to 1971 Census, with a range of 0 to 100 per cent, 100 among the Khasis in the north-east to almost zero amongst the Dorlas and Abujhmadias in Madhya Pradesh. A large number of tribal groups are still at about the 5–7 per cent literacy level. In many areas a few groups can hardly claim a university graduate.

The geo-climatic condition of the tribal regions is widely different, the only common feature being their hilly terrain and forest area. The forest may be very rich as in Bastar, or largely depleted as in Jhabua in Madhya Pradesh.

TRIBAL INTERESTS AND THE DEVELOPMENT OF FOREST ECONOMY: COMPATIBILITIES AND CONFLICTS

The nature of interdependence between the tribals and the forests and the interaction between the development objectives in the two sectors have been the subject of detailed studies, both locally and nationally, for a long time. The first significant study was that of the Scheduled Areas and Scheduled Tribes Commission, popularly known as the Dhebar Commission, in 1962. The other comprehensive study was by the Committee on Tribal Economy in Forest Areas (also known as the Hari Singh Committee) in 1967.

To promote educational and economic development of tribals, Tribal Development Agencies (TDAs), under the central government control were launched in eight areas. The economic programmes related to:

(1) land and water utilization, i.e. development of agriculture and horticulture, land development and soil conservation measures, water and power development, settlement of landless families, prevention of shifting cultivation, farm forestry and other afforestation measures;
(2) taking up a supporting programme of subsidiary occupation in a selective way;
(3) strengthening the co-operative and marketing infrastructure supplemented by a network of link and arterial roads; and
(4) undertaking debt-relief operations and safe-guarding against alienation of tribal lands.

In the Fifth Plan, planning for tribal development became a concurrent exercise with planning at state and national levels. Tribal sub-plan areas were carved out for areas with more than 50 per cent tribal population. Special plans were prepared for each identified sub-plan area. Thus, 65 per cent of the tribal population was covered under special sub-plans.

There is no doubt, however, that tribal economy and forest development are mutually dependent in most of the tribal areas. The tribal community traditionally has subsisted on forests and their produce where there is sufficient forest cover. It has been estimated by the Hari Singh Committee that roughly half the tribal area is covered with forests. Low productivity in agriculture in these areas has generally led to dependence of the tribal people on forests for their sustenance.

Since the economic, social and physical well-being of the tribals is so much dependent on forests, policy and programmes in the field of forestry are of immense significance. In a conference of Forest Ministers and Tribals Welfare Ministers of the states of Andhra Pradesh, Bihar, Gujarat, Madhya Pradesh, Maharashtra, Orissa, Rajasthan and West Bengal held in Delhi on 12 July 1978, the Prime Minister underlined the need for associating the tribals in bringing large areas under forest for restoring ecological balance by giving the tribals a right to the trees and their use in an assigned area.

Causes of disharmony

It has been stated that the forest administration has shown little concern for the tribals. Programmes for tribal development and forestry development have been formulated in isolation from each other.

The revised national forest policy introduced in 1952 (i.e. after Independence) curtailed the rights granted to forest-dwellers under the forest policy formulated in 1894. The new policy converted the rights of the tribals to forest resources into concessions. It (the new policy) recommended the classification of the forest into (1) protection forests, (2) national forests, (3) village forests and (4) treelands. The main goals in the management of the national forests are (i) to meet the requirements of defence, communication and industry; (ii) to wean the tribals away, by persuasion, from shifting cultivation; (iii) to control grazing in the forest, (iv) to promote the welfare of the people; and (v) to increase the effectiveness of forest management by providing legal authority through a specific set of laws. Thus the essence of the forest policy enunciated in 1952 was to promote state capitalism in the development and management of forests. The National Commission on Agriculture (NCA), recognizing the limitation of 1952 policy statement, has stated the revised goals as follows:

(i) to provide maximum goods and services for the well-being of the people and economic progress of the country;
(ii) to check denudation and erosion in mountainous regions and catchments of rivers;
(iii) to prevent erosion along treeless banks of rivers and on vast stretches of waste lands;
(iv) to maximize forest productivity with a view to meeting the growing demand for industrial raw materials, timber and other forest produce for defence, communication and domestic needs and to augment employment potential, with the ultimate aim of national prosperity;
(v) to provide for small timber and fuelwood requirements of the rural population;
(vi) to provide pasture for livestock in forest areas while ensuring against harm to the forests;
(vii) to provide recreational facilities in the forests;
(viii) to create blocks of forests interspersed with cultivation, or to introduce trees in larger numbers after careful selection, in order to maintain a healthy relationship between soil, vegetation and animal life;

There are a number of issues in the forest policy as it exists now:

(1) The conversion of right of the forest-dwellers into concessions would deprive them of their age old source of livelihood. Thus, this element is very punitive.

(2) It is doubtful if forestry will generate enough employment to compensate for the abridgement of the rights in respect to the use of forest resources by the forest dwellers.

(3) Although the NCA has emphasized the responsibility of the states to take concrete steps to ensure the supply of fuelwood, fodder, housing materials, and timber for village industries through social forestry and ancillary programmes, it is difficult to imagine that the states would meet their obligations in view of very limited amount of resources allotted for these programmes, even though the approved supply of fuelwood has only met 10 per cent of the demand.

In 1970, while the estimated requirement was 15.9 million m^3 (r) the production was only 8.92 million m^3 (r). In respect to fuelwood the estimated need was 175 million m^3, while the approved production was 13 million m^3. Thus, the production of fuelwood legally approved constituted 10 per cent of the estimated demand in 1980. One would expect a sizeable investment to meet this staggering gap, but actual allocation was very low. In the Fourth Plan 41.58 per cent of the total expenditure on forestry was invested in production forestry (excluding inventory planning and resource management, infrastructural development, development of mechanized logging, forest-based industries and so on). The investment for farm forestry and extension forestry including pasture development was about 5 per cent. About 4 per cent was spent for afforestation of degraded forest; but much of this included commercial forest species. Even when in the Fifth Plan mixed plantations in waste lands, Panchayat lands and forest areas, and also development of social forestry including reforestation of degraded forest, were treated as centrally sponsored schemes the situation did not improve. The expenditure on these two items during the first 4 years of the Fifth Plan constituted only 2.81 per cent and 3.56 per cent respectively of the total expenditure. Thus, notwithstanding strong sentiments expressed in favour of social forestry and allied programmes in the various official documents, the actual outlay is meagre. Further, during the Fifth Plan these programmes do not appear to have gained any momentum. They have stagnated more or less at the same level where they were during the Fourth Plan.

The other main areas of disharmony between tribal development and forest development are:

(1) Although the forest policy makes a mention of the local needs, it emphasizes commercial plantation for the wood industries. This has resulted in neglect of the tribals' needs such as fuelwood, housing materials, fodder, etc. This neglect has created considerable friction between the tribals and the forest administration.

(2) The scarcity of skilled manpower, needed for forest works which are labour-intensive, among the local tribals often has compelled the forest

department to bring the workforce from outside. The presence of outsiders, unfamiliar with local mores, customs and ethos, has created tension among the local population and the outsiders.

(3) The Forest Departments draw a large part of their manpower requirement from the tribals only as wage labour; and rarely provide a permanent support to the tribal economy.

The main solution to these discordant relations between the tribals and the forest development is to integrate the programme of the tribal sub-plans and the programmes for the development of forests.

Forestry programmes of relevance

One of the objectives of forestry planning is the generation of employment potential for the tribals. Forestry activities are labour-intensive, mostly employing unskilled labour. However, some amount of skill is required for a few activities, i.e. logging. Thus the labour-intensive activities, like making plantations and the construction of forest roads, which form the backbone of any forestry programme generate employment for tribals. Since plantations are likely to go on indefinitely on silvicultural and economic rotation, tribals in these areas are likely to continue to have income through these works. However, these works are unlikely to become permanent features of the tribal economy. Further, these activities will cover only a limited number among the tribal workforce. There is thus need for the development of forest-based programmes which form permanent elements in tribal economy.

In Gujarat the Forest Department has undertaken a pioneering programme to generate employment for tribals which may serve as a model for other states. Considering the importance attached to the preservation of tribal culture and the environment of the tribals, the main thrust of afforestation under various plans in the tribal areas has been on plantation, soil and moisture conservation, afforestation and social forestry works such as road-sides, canal banks, village woodlands, etc. In addition, the Forest Department has also recently introduced a few new schemes aimed at directly improving the conditions of the adivasis, thereby releasing them from the hold of the middlemen and moneylenders. Some of the schemes are briefly described.

Kotwalia scheme

Kotwalias, who are economically and socially the weakest of the tribals, sustain themselves mainly by manufacturing bamboo articles and selling them to the public. The literacy rate amongst them is only 1 per cent. Under the scheme the Government grants them bamboos at highly subsidized rates. Experience, however, showed that the subsidy granted by the Government did not ultimately benefit the Kotwalias despite putting in their hard labour for extraction of bamboos and manufacture of articles. This has happened because the middlemen advanced small loans to Kotwalia families at exorbitant rates of interest and in return appropriated the manufactured articles by paying very low prices and selling them at high rates in the market, earning huge profit. The end result was that the economic condition of the tribals did not improve in spite of the high subsidy given by the Government. The Department therefore conceived a scheme to eliminate the exploitation of the Kotwalias by the middleman under which Kotwalias are now supplied bamboos at their place of residence and are also given interest-free advance payments against the manufactured articles stored with the Department. These articles are then sold at competitive prices in major markets and the net proceeds are distributed to the Kotwalias after deducting the advances given to them and the royalty on bamboos and other expenditure incurred by the Department. The scheme has benefited the Kotwalia community where average earnings of families has increased substantially. In Dangs District, for example, 78 Kotwalia families have earned Rs. 211,000 since the inception of this scheme and in Surat District 400 families have earned Rs. 346,000 so far.

Planting trees in Malki lands

The adivasis of some of the backward areas of Dangs, Dharampur and other forests are not good cultivators. Most of their lands are on steep slopes in the hilly terrain with bouldery soils having very poor fertility. The edaphic conditions are also unsuitable for good and remunerative agriculture. The agricultural returns from such areas are not more than Rs. 100 per hectare per year. Often, due to floods or scarcity conditions, crops fail. The present scheme therefore envisages that in the Malki lands, now under marginal cultivation, it is possible to raise trees as an economic crop. The tree crop is expected to give higher financial returns at the end of the very short rotation of 15 years. The forest crop will also conserve the valuable soil in this hilly tract which otherwise gets eroded due to unscientific cultivation and inadequate soil-conservation measures. The scheme will therefore not only benefit the malki owner but also protect the precious soil from erosion.

An adivasi, who is willing to take advantage of this scheme, has to hand over one or two hectares of his land to the forest department for raising teak and bamboo plantations. He is given a subsistence allowance of Rs. 250 per hectare per year which is equivalent to his expected average annual earnings from cultivation of such land till the time of exploitation. At the end of a 15-year period, teak trees will be exploited by the Forest Department and sold on behalf of the landholder. From the gross proceeds it is proposed to deduct the cost of the plantation and also the subsistence allowance paid to the landholder over the intervening years without charging any interest, and hand over the balance amount to the landholder. In Dangs District, 64 families of adivasis have so far taken advantage of this scheme and plantations have been raised over 100 ha of their lands.

Establishment of grain banks in forest areas

The economic condition of the adivasis, dwelling in forest areas, is very precarious; hence they become victims of economic exploitation by shopkeepers and moneylenders. In the monsoon season the adivasis are mostly unemployed and have no money to purchase the food grains. They procure them from the local moneylenders and shopkeepers on loan. When their new crop comes in they return the loan, mostly in the form of grains or oil seeds, the value of which is about 3 to 4 times more than the actual grain or money taken on loan. In order to prevent this type of exploitation of the adivasis, the Forest Department has established food grain banks in forest villages, where the adivasis can get food grains on loan in summer and monsoon when they have no food grains in stock. Each adivasi family is given food grains twice or thrice in a season, as per their requirement. After the monsoon season, when the crop is harvested, the quantity of food grains equivalent to what had been taken on loan, together with a 10 per cent surplus to compensate for packing and handling charges, is to be returned by the adivasi cultivator to the bank. The food grains thus received are stored in the grain bank and the same are given on loan in the following year. Food grain banks have been established in Bharuch, Surat and Valsad Districts and this scheme is being extended to all adivasi areas of the state. In 1977–78, 17 food grain godowns were constructed and grain purchased at a total cost of Rs. 823,000. During 1978–79, 18 more food grain godowns were constructed and grains purchased at a cost of Rs. 995,000.

Social security for tribal families through afforestation

In some of the tribal areas of Panchamahals and Bharuch the tribals are mostly landless and without regular employment. These tribals migrate to far-off places every year, in search of manual work; this results in an unstable social life for them. The Forest Department has therefore developed a concept under which it is proposed to provide an assured monthly income by permanently engaging a family of adivasis on forest plantation works. An adivasi family will be earmarked an area of 2.5 ha of land each year for raising plantations. Plantations of suitable forest species will be raised in this land by the family under the direct supervision and guidance of the Forest Department. For raising plants, the required material, i.e. polythene bags, seeds, etc. will be supplied by the Department. The family will be responsible for all operations—digging of pits, soil conservation measures, planting, weeding, fencing and protection for successful raising of the plantation. Each family will be paid a monthly remuneration of Rs. 250 over a period of 15 years. The family will also be given small timber, bamboos, etc. for constructing a temporary hut near the plantation area. At the end of the rotation period of 15 years the family will be given a 20 per cent share in the net profits derived from the sale of the material, making them partners in the profits. This measure, it is hoped, will help to stabilize the families, and in the process ensure better protection of forest, and also make available fuelwood and small timber which is in great demand. The economic and social conditions of landless adivasis in the state will also improve.

It is a national forest policy to set apart a substantial part of the forests to produce raw material used in the production of the goods of mass consumption like matches, pulp and paper and hence to undertake production forestry in all suitable forest areas including those inhabited by the tribals. This will mean more intensive forestry operations which may also include clear-fellings and replanting with species of industrial and economic value. However, it is possible to build some safeguards in this policy which will enable the tribals to derive their traditional benefits from the forests.

A new programme in the forestry sector has been taken up recently on the recommendations of the National Commission on Agriculture on Social Forestry, which aims at creating as much tree cover in wasteland and degraded forest areas as possible. Social forestry is expected to contribute to substantial employment in the rural areas. It is also expected to produce small timber, fuel and fodder, etc., which are basic needs in rural areas. Social forestry is also linked with other aspects of rural development. For example, in the choice of species, consideration should be given to such factors as needs of fodder for livestock, foliage for tassar, muga, lac, silk cultivation, etc.

Minor forest products play a very important role in the tribal economy, particularly in more backward areas. The rights of tribals to minor forest products for bona fide domestic consumption, and in some cases for sale, were recognized while the forests were reserved. With the change in socio-economic conditions of the tribals, the dependence on these minor products has been reduced. They now provide for small cash needs of the tribals. Collection of these products has been carried on for quite a long period on a modest scale, through contractors. But at present with the large appreciation in value, and also importance of some items of the minor products in the national economy, the states are commercializing the operations. This has inevitably led to some friction between the rights of the tribals and those of the states. Fresh thinking is needed to resolve this conflict.

RECOMMENDATIONS FOR HARMONIZING TRIBAL AND FOREST DEVELOPMENT PLANS

Production forestry, which involves a large-scale afforestation, seems suitable to creating harmony between forest development and tribal development.

Afforestation is a major part of production forestry congenial to the tribal way of life. However, plantation of commercial trees may not appeal to tribal communities and it may be necessary to assure them that the development of forests will really strengthen their economy, besides being in the interest of regional and national development. The projects for production forestry should be prepared after fully considering their implications for the local tribal economy. A programme of intensive exploitation of forests often creates demand for increased manpower. In the same area where production forestry is planned, an intensive agricultural development programme may also be taken up which would require more labour, providing opportunities for employment to tribals. There is a danger in this approach which should be recognized. These programmes may compete for locally scarce manpower and may lead to an influx of outside labour, which may not be in the interest of socio-economic balance nor in the preservation of forests. The following approach may bring about greater harmony between the demands of intensive development and the interests of the local tribes.

(1) Programmes of training for developing the necessary skills among the tribals needed to carry out new programmes (for instance training in manual and mechanized logging) should be a part of the commercial forestry. This will reduce the need for outside labour.
(2) In raising plantations for industrial and commercial use a judicious mixture of species which may yield fodder, fruit or minor products should be introduced. Similarly while clear-felling areas for plantation, species which have been traditionally linked with the tribal economy and are of daily use to the tribals, should be preserved.
(3) A greater emphasis should be placed on education and social welfare services in tribal areas which would enable the tribals to cope with changes in tribal economy.

Minor forest products will continue to play a key role in the harmonious relationship between the forest and tribal development. It is recognized that certain items such as tendu leaf and sal seeds should be developed for commercial use only. But a principle should be accepted by the states that instead of considering the tribals who are engaged in collection of such produce as mere wage labourers, a remunerative price should be arranged for the material to be collected. Thus one of the important forest-based planned activities would be for the Forest Departments either directly or through tribal cooperatives to organize collection and marketing of these products, in order to ensure remunerative prices for the benefit of the tribals.

Social forestry should be recognized as one of the priority forest programmes in the tribal sub-plan areas. The species grown in the forests should be such as can be alloted to tribal families who will manage them without transferring rights to land. They may be compensated partly in cash and partly in kind. The examples are fodder trees, suitable lac trees, tassar and muga cultivation. Similarly, fuel plantations can be planted with which identified tribal populations can be linked.

The tribal sub-plan programmes in sectors other than forest development include construction of roads, provision of water supply, building of educational institutions, etc. For quite a long time the Forest Departments has provided for these services under their own plans for 'tribal villages' in the forest areas. The Forest Department personnel are also trained to undertake such works. But the image of the Forest Department is of a group of individuals trying to restrict the rights of the tribals while the other Departments are viewed as supporters of the tribals. It may be suggested that the welfare activities belonging to other sectors in sub-plans, and which can be implemented by the Forest Department, should be entrusted to them. This will not only reduce the overhead but will use the existing machinery to implement welfare programmes.

The trust of the tribal has to be won through imaginative implementation of the forest policy. This requires different orientation for which the Forest Departments will have to prepare themselves. So a cadre of forest officers and other field personnel, properly trained and wedded to the principles of tribal development, would have to be built in the Forest Departments and deployed in tribal areas. As most of the tribal sub-plans have more than 50 per cent of the area under forests, and the main thrust of the integrated plan would be based on forest, selected forest officers should be appointed as project administrators in the integrated tribal development projects.

There is a constitutional obligation of the central government for the development of the tribals. Consequently the centre has to take the lead in ensuring the implementation of the programme. For this purpose, a special sub-cadre for looking after harmonious tribal and forest development plans should be built up both at the centre and the states.

It should be noted that tribal villages in the forest areas are not covered by tribal sub-plans. The system of forest village should be done away with, and should be replaced by a contractual obligation for payment of reasonable wages to which tribal families work for the Forest Department. This was also the recommendation of the Forest and Tribal Welfare Ministers made in a Conference held in July 1978 at New Delhi. However, till these villages are converted into revenue villages the Forest Departments should endeavour to take up a programme based on individual families rather than on the community.

Socio-economic Effects and Constraints in Tropical Forest Management
Edited by E. G. Hallsworth

5

Socio-economic effects and constraints in forest management: Tanzania

BJORN LUNDGREN
International Council for Research in Agroforestry (ICRAF), Nairobi, Kenya
and
LILL LUNDGREN
Department of Physical Geography, University of Stockholm, Sweden

INTRODUCTION

Tanzania is situated on the east coast of Africa, roughly between latitudes 1° and 12°S. It occupies an area of 945,000 km^2 of which 55,000 is inland water (Lakes Victoria, Tanganyika, Nyasa and Rukwa). Physiographically, the country can be divided into:

(1) *the coast region,* made up of a narrow zone of coastal coral platforms and sediments and a low dissected plateaux hinterland, mostly composed of resistant layers of Jurasic and Cretaceous sandstones;
(2) *the highland plateaux regions;* these cover by far the largest area of the country, being mainly vast inselberg plains; they are made up of various crystalline rocks; the altitude is normally between 500 and 1500 m;
(3) *the Rift Valley volcanic highlands,* separated in Tanzania into two distinct areas—a northern one associated with the Gregory Rift including the volcanic peaks of Mt Meru and Kilimanjaro, the highest peak in Africa (5895 m), and a southern one along Lakes Tanganyika, Rukwa and Nyasa;
(4) *block mountain massifs* formed through faulting and uplifting on the crystalline peneplains; they often rise steeply up to around 2000 m altitude; the most important ones are the Usambara Mts, the Uluguru Mts, and the Southern Highlands (including Sao Hill).

The vast interior high plains are dry or very dry with an annual rainfall varying from 400 to 1000 mm, falling normally in one season with a severe and prolonged dry season. Higher rainfall is encountered on the coast, in the Lake Victoria basin and in association with the uplifted mountain areas and volcanic peaks, where orographic precipitation up to 2000 mm is common. In the most northern parts of the country, close to the equator, the rain falls in two seasons. Temperatures are uniformly high throughout the year except in the higher mountain ranges where temperatures may be very low and frosts occur.

The crystalline plateaux highland soil catenas are dominated by weathered and leached ferralitic and ferruginous soils with various alluvial, vertisolic and calcic soils developed in areas of specific topography, hydrology, geology and climate. Andosols of various ages and degree of development are found in the volcanic highlands and mountains, whereas the block mountain massifs have mixed lithosolic and ferrisolic soils.

The distribution of the vegetation is predominantly governed by climate, soil and human activities. The drier parts of northern and central Tanzania are covered by savanna communities with various degree of tree and bush cover (bushland, wooded/bushed grassland, and grassland). In the western and south-eastern plateau country the most widespread vegetation type is 'miombo' woodland, which is a deciduous forest type with a light canopy, dominated by leguminous trees (*Brachystegia, Julbernardia* and *Isoberlinia* spp.). It is often associated with *Pterocarpus angolensis*, one of Tanzania's major commercial timbers. A very small part of the country, less than 2 per cent, or about 15,000 km^2, is covered by closed evergreen forest communities. Most of these are found in the mountains. Very limited areas of lowland and intermediate rainforests remain. In appearance the montane forest is similar to that of lowland forest except for a smaller number of buttressed trees and fewer of the great giants. The best known and most useful dominant trees are *Podocarpus* spp. (conifer) and *Ocotea usambarensis* (East African camphor). In the drier ranges of the mountains another valuable conifer, *Juniperus procera* (East African pencil cedar), is common. Dense thickets of mountain bamboo (*Arundinaria alpina*) occur on some mountains.

POPULATION AND LAND-USE

In 1978 the total population of Tanzania was 17.5 million with an annual growth of 3 per cent. Most people, 87 per cent, live in rural areas and the remaining 13 per cent in towns. The average population density was 19 persons per km^2. The distribution of the rural population is uneven with strong concentrations in some areas, contrasting with large parts of the country which have very low densities. More than 75 per cent of the population concentrate in less than 15 per cent of the total land area. The main regions with dense population are near the coast, in the vicinity of the great lakes (particularly Lake Victoria), and in the mountainous areas, a distribution that coincides well with areas of high rainfall (Nieuwolf, 1977).

Other factors which exert an influence on population distribution are the occurrence of tsetse flies which, in combination with poor soils and lack of water, makes the vast tracts of miombo woodlands very sparsely populated in spite of comparatively high rainfall. Also historical events have partly influenced the present population distribution, particularly tribal wars, migrations and slave trade during the eighteenth and nineteenth centuries.

Since the late 1960s a considerable local redistribution and concentration of the rural population has occurred as a result of the so-called 'villagization programme'. Among the aims of this programme has been to render social services to the rural population, which was difficult when people lived in scattered homesteads over vast areas. Another has been to facilitate more efficient introduction of improved farming techniques. The enormous magnitude of this programme is evidenced by the fact that, since 1973, 12 million people were moved into 8000 villages (Mnzava, 1980).

In Tanzania all land is owned by the state, although there are still some larger cash-crop estates (coffee, tea, wattle, sisal) held under long-term leases. The total area of these estates is insignificant but they are normally situated on high-potential land. In some parts of the country, particularly in wet upland, high-potential areas, land is still *de facto* regarded as private and is inherited between generations (Mnzava, 1980). The fragmentation of such land has reached a critical stage in many upland areas (e.g. Kilimanjaro, Meru, Usambara, Uluguru, Mbeya) a factor that strongly contributes to the pressure on adjacent forest reserves.

In the drier and more marginal land shifting cultivation was until recently the traditional form of land-use. The villagization programme, in combination with certain legal measures to prevent people from reverting to traditional forms of living, has led to a considerable decrease in shifting cultivation over the last decade. In many places the result has been a serious pressure on land resources near the villages. Trees have been cut in large circles around the villages, both through clearings for agriculture and for fuelwood. Already after a few years many newly established villages suffer from erosion and declining land productivity.

It has been estimated that only around 5 per cent of the total land area is under small-scale agriculture (large-scale less than 1 per cent). The major staple food crops are maize, bananas and beans in the wetter areas; and maize sorghum, millet and cassava in the drier areas. Among the more important smallholder cash crops are coffee, tea, pyrethrum and vegetables in the mountains; and cotton, tobacco and groundnuts on the arable plains. Along the coast there is a high production of cashew nuts. The vast savanna areas, almost 50 per cent of the total land area, are mostly used for extensive grazing, where they are not set aside as national parks of game reserves. Except for some expanses of swamps the remaining area, or roughly 40 per cent of the country, is classified as forest and woodland (Berry, 1971).

FOREST LAND-USE

The pre-colonial history of forest land-use in Tanzania is not well known. Shifting cultivation was practised in the woodland and moist savanna areas, but due to the sparse population the effect on the vegetation was probably slight. In the miombo woodland the occurrence of tsetse flies prevented large-scale habitation, and hunting and honey-collection were the main forms of utilization. In the drier areas, however, the influence of pastoralists on the tree vegetation, through grazing and the systematic use of fire, was certainly much more profound.

There are strong indications that mountain forest areas were cleared for cultivation during earlier dry periods, e.g. around A.D. 1000, and then kept open by fires and grazing (Lundgren and Lundgren, 1972). The mountain forests were often also used as temporary refuges during times of war and it is likely that a few permanent settlements have always existed in the mountain forests. Large-scale movements of population up the mountain slopes occurred first during the eighteenth century, primarily as a result of the invasion and occupation of the plains by the cattle-raising Masai. This forced the agriculturalist Bantu tribes to settle in the mountains where they could better defend themselves. They cleared the forest on the lower slopes and, depending on skill and land potential, different forms of permanent and semi-permanent agriculture developed. Some of these mountain tribes developed into powerful and well-organized societies during the eighteenth and nineteenth centuries, e.g. the Chaggas of Kilimanjaro and the Sambaa of Usambaras.

When the first Europeans came to Tanzania in the late nineteenth century, forests were still widespread in the mountains and in the wet parts of the country. The country became a German colony in the late 1890s and in 1897 organized forestry activities had already started. The main priority was protection and reservation of natural forest, with a dual aim of protecting the water catchments and securing production of timber. Up to 1914, 231 forest reserves were delineated with a total area of more than 750,000 ha. The reserves were mainly montane and mangrove forest. Although some people were probably evicted from the forests during this reservation, as evidenced in the occurrence of secondary forests within the reserves, it is likely that most forests brought under protection were uninhabited (B. Lundgren, 1978).

During the German time some sawnmills were erected for exploitation of the valuable montane trees. Plantation trials, both of indigenous and introduced trees—teak, for example, was brought into the country in 1898—were initiated, but plantation forestry never achieved importance (only 700–800 ha were planted before the outbreak of war (Grant, 1924).

Forestry activities were resumed again in the early 1920s after a complete standstill during the war, when a considerable encroachment took place into the reserves by people trying to escape enlistment into the German colonial army. The first task for the new British administration was therefore to reclaim the old montane forest reserves. In the 1920s and 1930s reservation was also made for large tracts of miombo woodland to secure the supply of the valuable *Pterocarpus angolensis*. At the end of the British time the areas under forest reserves were practically the same as today. The total area of reserves is 130,000 km^2, of which only 9300 km^2 are high forests and the bulk of the remainder is miombo woodland (Parry, 1962; B. Lundgren, 1978).

The major forestry development during the British time was the systematic introduction and trial of exotic trees suited for plantation production of timber, firewood, and later pulpwood. Large-scale planting started in the early 1950s, mainly in highland forest reserves. The major species were pines, cypresses and eucalypts. This plantation activity has continued and accelerated after the country became independent in 1961. Today there are some 70,000 ha of fast-growing industrial plantations, with an additional 4000 ha being established every year. A major share of Tanzania's industrial wood need is now met from these plantations. The balance still comes from logging in the natural forests, mainly the miombo woodlands (B. Lundgren, 1978).

Forest policy and legislation are practically the same today as they were laid out in colonial times. Emphasis is laid on protecting and managing the reserved forest in order to achieve perpetual production of wood for the country's need and to secure ecological and hydrological stability of major watersheds. People are not allowed to reside in forest reserves or to use any products from them, although locally customary rights have developed, whereby people residing near the reserves are allowed to collect branches as firewood or to herd cattle through the reserves along special tracks. The Forest Department and its staff have, in the eyes of the rural people, often been regarded as representatives of higher authorities with a policing task to prevent people from using land and wood resources. Only during the last few years, in connection with the villagization programme, has the Forest Department taken a more active part in educational and extension efforts, aiming at informing people on the benefits of forests and trees. The policy is currently undergoing some revision, which will concentrate on forestry not only for, but also by, the rural communities (Mnzava, 1980).

There has, however, been one significant activity whereby people have benefited in the recent past from forestry in reserved areas, apart from direct employment opportunities in forestry and forest industries. The 'taungya' afforestation system has been widely used in establishing most of the presently existing highland plantations. It was first used by the Germans but it was not until the 1950s, with the expansion of plantation establishment, that it began to exert an important socio-economic influence in the densely populated mountain areas. In this system the forester assigns pieces of natural forest land to licensed cultivators, normally 2–5 ha per family. The farmer clears the forest (after valuable trees have been extracted by the Forest Department), burns the slash and cultivates his crops, normally maize, beans, peas, potatoes (Irish), and pyrethrum (only annual crops are permitted) for 3–4 years. In the first or second year the Forest Department plants seedlings at regular intervals in the farm, and these are thus intercropped with the farmer's crops for 2–3 years when the canopy closes and the farmer is allotted a new piece of forest. It is the farmer's responsibility to tend the young trees (weeding, protecting them from animals, etc.), while he cultivates the land in between. The benefit to the Forest Department is obvious–land clearing and weeding obtained at no cost.

During the last few years the system has, however, been discouraged on social (temporary houses, poor service to the farmers), political (making the villagization programme difficult etc.), ecological (natural forests should be preserved), and forestry (people damage trees and refuse to move from farms) grounds. Still, it has been shown that the economics of the system are beneficial, not only to the Forest Department, but also the farmer (Hofstad, 1978). The reason is that mountain forest land is the most fertile land available in the country and a good farmer can obtain better harvests on a 'taungya' plot than in most other places.

The exact figure for fuelwood consumption in the country is not known. Estimates vary between 20 million m^3 (r) (Gilliusson, 1976) and 35 million m^3 (Mnzava, 1980), an indication of the difficulty of obtaining reliable information. This represents between 85 and 95 per cent of the total wood consumption. Most of this wood, probably more than 90 per cent, comes from unreserved forests, woodlands and savannas. It is safe to assume that the local and regional rate of increase in fuelwood consumption has been very high in the last decades. There are three main reasons for this. The most apparent one is the increased population. Secondly, the concentration of people in villages has resulted in large local deforestation as a result of clearing forest for cultivation, collection of fuelwood and the momentary need for large amounts of construction timber and poles when erecting the villages. The third reason is the expansion in certain regions of tobacco-growing and curing which requires large quantities of fuelwood. In Tabora Region, the main tobacco-producing region, 1.7 million m^3 of fuelwood was used in the 1977/78 season for curing the tobacco. It is estimated that the 1980/81 tobacco production target in the region will require the clear-felling of 35,000 ha of miombo woodland just to meet the need of fuelwood for curing tobacco (Temu, 1979).

There is little doubt that the rate of deforestation is rapidly increasing. The often-quoted figure of 300,000 km^2 being under unreserved forest (mainly miombo woodland) does certainly not hold true any longer. In the semi-arid areas the situation is already very serious with erosion and other land degradation problems, as well as serious socio-economic implications for the rural population. As wood becomes scarce fuelwood collection for the household becomes a burden and progressively more labour is diverted, especially for women. Health hazards as a result of bad cooking and fatigue increase.

THE MOUNTAIN AREAS

The highlands of Tanzania have a considerably higher potential for crop production than any other biogeographic region of the country (Lundgren, L. and Lundgren, B. 1979). This fact is due to a combination of favourable rainfall, soil and photosynthesis conditions. The mean rainfall is adequate for rain-fed cultivation of most crops. Soils generally have a high cation-exchange capacity, a high base saturation and a higher organic matter content than soils of lower altitudes. This is valid also for the mineralogically rather poor soils derived from basement rocks, e.g. those of the Usambara and Uluguru mountains. Finally, the potential photosynthesis is higher due to the lower respiration rates resulting from the lower temperatures. Estimates have been made that the highland – lowland difference in potential photosynthesis give a 2 to 1 (possibly even greater) advantage of the highland areas over the lowland areas with respect to grain yields (Porter, 1979).

These factors have resulted in high population densities. Over the last 80 years the population in the West Usambara Mts has, for example, increased from an estimated 15,000 in 1900 to 286,000 in 1978 (L. Lundgren, 1980). As a result, deforestation has taken place on a large scale on the mountain slopes. The impact of land-use practices—predominantly small-scale arable farming—has become progressively more intensive. This over-cultivation has, in turn, led to severe soil erosion and land degradation in many mountain areas. Erosion and fertility decline are obvious limiting factors to land utilization in these regions. Different types of erosion hazard are important on deforested and over-cultivated tropical mountain slopes, viz. splash erosion and rill erosion occurring every year and landslides triggered off in years of extreme rainfall.

It must also be borne in mind that vast areas surrounding the mountains are dependent on the water quality and flow characteristics of the permanent rivers that rise in the highlands. Finding forms of land-use which will give the best yields of food and cash crops on a sustained basis, while at the same time preventing erosion and maintaining a favourable hydrological balance, is an urgent need.

In the 1930s the deterioration was already obvious and the British administration launched soil conservation and rehabilitation schemes. Important schemes in this area were the Mlalo Basin Rehabilitation Scheme and the Usambara Development Scheme in the Usambara Mts, and the Uluguru Land-Use Scheme in the Uluguru Mts, all being conducted in the post-war time. Both the Usambara and the Uluguru Schemes were unsuccessful and had to be given up in the middle 1950s. The failure can be attributed to unsound and unwise implementation and practices that were badly adapted to traditional social and cultural systems. In addition, growing unrest as independence approached contributed to resistance towards the Colonial Administration, including the conservation schemes (L. Lundgren, 1978, 1980; Temple, 1972; Watson, 1972).

Another result of this development—increasing population and land deterioration—is that pressure on the forest reserves is considerable. In 1963 politicians yielded to this pressure and excised 12,000 ha of a forest reserve in the Usambara Mts. Today this area is totally deforested, also on the steepest slopes, and erosion is very severe, indicating the speed of land degradation. A smaller area of 500 ha, already planted with cypress and pines, had to be given up 1971–72 by the Forest Department after the taugya farmers refused to leave the area (B. Lundgren, 1978). Another example is from another part of the Usambara Mts. In this area the natural montane forest of a catchment belonged to two different tea estates, bordering on forest reserves. As one of the estates was nationalized, the forest—being one of the most luxuriant in Tanzania and unique in connection with the rest of the catchment forest—was opened for local cultivators in the early 1970s. The other part, previously owned by the other estate, was donated to the University of Dar es Salaam just before the pressure for releasing it for local cultivation arose. This University Forest Reserve (Mazumbai), although small (400 ha), is one of the most important baseline research areas in East Africa today.

It is of fundamental importance that the changes and intensifications in land-use practice which occur as a result of the rapid population increase are based on evaluation of interactions between land management, vegetation (including crops), soils and climate, if the full land potential of the highlands is to be realized. Misjudgements or ignorance may easily have devastating results, not only on the highlands themselves but also on its surrounding low-lands which depend on the mountains for their water supply.

THE SITUATION TODAY

In summary, the following issues are relevant when evaluating the socio-economic effects and constraints in forest management in Tanzania today:

(1) The villagization programme has in a short time created enormous concentrated pressures on forest and wood resources throughout the coun-

try, but particularly in the drier areas on the high plains and in the Lake Victoria area, where deforestation was severe even before villagization.

(2) The dense and rapidly increasing population in the high-potential mountain areas exerts a considerable pressure on land resources and the continued protection of the essential catchment forest reserves is endangered.

(3) In some regions the rapid expansion of agricultural uses such as tobacco-growing has resulted in a very rapid large-scale deforestation.

The Tanzanian Government and Forest Department are aware of these problems and over the last few years considerable emphasis has been given to develop village forestry programmes. The aim is to make the new villages self-supplying in wood on a sustained-yield basis by establishing village woodlots. The Forest Department has increased its extension efforts by producing plants free of charge, education programmes via primary schools, and tree-planting campaigns. Resources are, however, quite insufficient to keep pace with the rapid development of this enormous problem.

The rapidly deteriorating situation in some mountain areas is not fully appreciated yet. These problems are complicated and their solution, or at least the prevention of further deterioration, will require integrated efforts spanning over geographical, administrative, departmental and other boundaries.

Research is going on, although on a small scale, to find management methods that will secure a sustained supply of curing fuel from the miombo woodland and from fuelwood plantations for the tobacco-growing villages (Temu, 1979).

REFERENCES

Berry, L. (ed.) (1971) *Tanzania in Maps*. University of London Press. London.

Gilliusson, R., (1976). Outlines of forestry projects in Mwanza and Shinyanga Regions to be presented for financing by the World Bank. Cyclostyled. Min. Nat. Res. and Tourism. For. Div. Dar es Salaam.

Grant, D. K. S. (1924) Forestry in Tanganyika. *Emp. For.*, **3**, 33–8.

Hofstad, O. (1978). Preliminary evaluation of the taungya system for combined wood and food production in north-eastern Tanzania. Rec. no. 2. Div. For. Faculty Agric., For. Vet. Sci. Univ. Dar es Salaam.

Lundgren, B. (1978). Soil conditions and nutrient cycling under natural and plantation forests in Tanzanian highlands. Reports in forest ecology and forest soils. Dep. For. Soils, Swedish Univ. Agric. Sci. No 31, Uppsala.

Lundgren, L. (1978). Studies of soil and vegetation development on fresh landslide scars in the Mgeta Valley, Western Uluguru Mountains, Tanzania. *Geogr. Ann.* **60A** (3–4) 91–127.

Lundgren, L. (1980). Comparison of surface runoff and soil loss from runoff plots in forest and small-scale agriculture in the Usambara Mts., Tanzania. *Geogr. Ann.* **62A** (3–4), 113–48.

Lundgren, B., and Lundgren, L. (1972). Comparison of some soil properties in one forest and two grassland ecosystems on Mount Meru, Tanzania. *Geogr. Ann.* **54A** (3–4), 227–40.

Lundgren, L., and Lundgren, B. (1979). Rainfall, interception and evaporation in the Mazumbai Forest Reserve, West Usambara Mts., Tanzania, and their importance in the assessment of land potential. *Geogr. Ann.*, **61A** (3–4), 157–78.

Mnzava, E. M. (1980). *Village Afforestation Lessons of Experience in Tanzania.* FAO TF/INT 271 (SWE).

Morgan, W. T. W. (1973). *East Africa.* Longman, London.

Nieuwolt, S. (1977). The influence of rainfall on rural population distribution in Tanzania. *J. Trop. Geogr.*, **44**, 43–56.

Parry, M. S. (1962). Progress in the protection of streamsource areas in Tanganyika. *E. Afr. Agric. J.*, **27**, 104–6. Special issue.

Porter, P. W. (1979). Potential photosynthesis and agriculture in Tanzania. *BRALUP Res. Paper* **29**: 1–59 and Paper **29**(1): 1–71.

Temple, P. H. (1972). Soil and water conservation policies in the Uluguru mountains, Tanzania. *Geogr. Ann.*, **54A** (3–4), 110–23.

Temu, A. B. (1979). Fuelwood scarcity and other problems associated with tobacco production in Tobora Region, Tanzania. Rec. No. 12. Div. For. Faculty Agric., For. Ver. Sci. Univ. Dar es Salaam.

Watson, J. R. (1972). Conservation problems, policies and the origin of the Mlalo Basin rehabilitation scheme, Usambara Mountains, Tanzania. *Geogr. Ann.*, **54A** (3–4), 221–6.

Socio-economic Effects and Constraints in Tropical Forest Management
Edited by E. G. Hallsworth

6

Problems of shifting cultivation in north-eastern India

E. S. THANGAM
Chief Conservator of Forests, Arunachal Pradesh, India

INTRODUCTION

The practice of shifting cultivation has been traced back to 7000 B.C. according to archaeological data (Sharma, 1976). It was in existence throughout the world, including the Danube Valley and Formosa (Childe, 1976). This method was adopted when man changed from the food-gathering to food-production habits. Even now this form of cultivation is practised in most tropical countries in Africa, South America, Oceania and South-east Asia. It is estimated that an area of about 36 million km^2, or about 30 per cent of the world's exploitable soils, are at present under shifting cultivation, producing the bulk of the food for more than 250 million people or 8 per cent of the world's population. In India about 3 million people are engaged over an area of 5 million ha in various forms in fifteen states, viz. Andhra Pradesh, Arunachal Pradesh, Assam, Bihar, Karnataka, Kerala, Madhya Pradesh, Maharashtra, Manipur, Meghalaya, Mizoram, Nagaland, Orissa, Sikkim and Tripura.

The expression 'shifting cultivation' covers a large variety of traditional forms of agriculture practised by primitive and other people in the hills. The practice consists in clearing the forests, setting fire to the felled material after it becomes dry and raising agricultural crops for one or more years, followed by a fallow period. The cycle is repeated at the end of a fallow period. It is also known as 'shag', 'slash-and-burn agriculture', 'jhum', 'swidden', etc.

Conklin (1961) described shifting cultivation as 'any continuing agricultural system in which any impermanent clearings are cropped for shorter periods in years, then are fallowed' Bene, Beall and Cote (1977) described it as 'a sequential system of growing trees and food crops'.

This cultivation system was well adapted to the environment under certain socio-economic conditions of the tribal people who practised it in a traditional manner. The traditional customs, and the knowledge of the people for growing both cereal and tree crops, catered for the maintenance of the ecological balance. However, this way of life was maintainable only so long as the people's numbers remained low and they were unable to find a solution to raise the long-term carrying capacity of forest lands when the population increased. Since 1950 the population growth has accelerated and the increased number, in turn, has triggered off land hunger and started migration to the hills. The pressure from the settled cultivators and 'land sharks', or speculators, aggravated the problem. There were also large and growing throngs of squatters who, finding themselves landless, turned to forest lands as an agricultural opportunity of last resort. As a result, the forest cover in the hills is being stripped off at an increased rate at every available opportunity. Conway, Gordon and Romm (1972) put it clearly:

> The forests of tropical South-east Asia are among the richest and most productive in the world. This impression of great natural wealth can be deceptive, however, for the productivity depends on a fairly limited and fragile source of nutrients. The tropical forest maintains itself almost entirely on nutrients in the surface layers of the soil. At any given time, a high proportion of the ecosystem's nutrients are bound in its vegetation rather than in its soil. Continued soil fertility depends on the return of vegetative detritus to the surface and the assimilation of the litter breakdown products by the soil. Unless careful consideration is given to this nutrient cycle when trees are removed, the thin surface layers of the soil will lose their fertility and erode, leaving the much poorer sub-soil. Crops can be planted to replace forest growth. They will thrive for a while if the surface layers of the soil have been retained but if the nutrients are not replaced, yields will gradually decline.

Due to the variations in topography, rainfall, land tenure, agricultural heritage and temperature, the pattern of shifting cultivation is not uniform all over the country. Further, the intensity of shifting cultivation varies with the changing conditions of rainfall, topography, density of population and accessibility.

Shifting cultivation is generally practised in India in the tropical wet evergreen, semi-evergreen, moist deciduous and dry deciduous forest, subtropical broadleaved and pine forests and the Himalayan moist temperate forest types, up to an elevation of about 2000 m.

On the extent of dependence, shifting cultivators could be classified as those primarily dependent, partially dependent and marginally dependent on shifting cultivation. The extent of dependence on shifting cultivation by a tribal community is directly proportional to the distance of the area from the growing urban centres. The tribal people who live close to the urban areas still practice shifting cultivation but only as a traditional custom, thus depending marginally on it. People who live further away depend partly on shifting cultivation as they also take up various jobs, including labour work on daily wages from time to time. In the more remote areas, due to lack of infrastructure development, the tribal people depend primarily on shifting cultivation.

The north-east region consists of seven states, viz. Arunachal Pradesh, Assam, Manipur, Meghalaya, Mizoram, Nagaland and Tripura. According to the 1971 census, the population of the region was 19.58 million, with 90.56 per cent of the population living in rural areas, as against 80.09 per cent for India as a whole. The per capita land available for cultivation in the north-eastern region is 0.13 ha, including cultivated land and fallow land, as against 0.26 ha for India as a whole in 1972–73. Due to the hilly nature of the terrain, a smaller area is available for settled cultivation, particularly with increased population.

THE PROBLEM

In the pre-independence period, two main approaches were adopted towards the tribals, viz. (1) total isolation by segregation of tribal areas; (2) detribalization for assimilation among others. The first approach created pockets of tribal society leading their lives in their own way, and the second made them citizens in the lowest range of the society. After Independence, the policy adopted was given in the following words of Jawaharlal Nehru: 'We do not mean to interfere with their way of life, but want to help them to live it . . . according to their own genius and tradition.'

Legislative measures likle Jhum land Regulation, 1948 of Assam were not successful as it covered only limited areas and was difficult to enforce. The National Forest Policy, 1952, which laid stress on weaning away shifting cultivators by persuasion, could not achieve any perceptible success. The taungya system, which succeeded in West Bengal, was not successful in other states as it was unable to mobilize all the shifting cultivators and was confined to certain groups, particularly landless and lowland migrants. Intensive methods of agriculture as an alternative to shifting cultivation also did not prove successful as they failed to cater for the needs of all the shifting cultivators and were not suited to the socio-economic life of the tribal people. Since the Fourth Five-Year Plan, with emphasis on micro-level planning, the tribal sub-plan area programmes, catering for the special needs of the tribal people with an integrated area development approach having special funds, were introduced with success. Other programmes were also taken up by the state and central government like the 'Pilot Project on Control of Shifting Cultivation' in eleven states in 1976–77 and the Small Marginal Farmers Development Agency during the Fifth Plan (Thangam, 1979).

The North-Eastern Council, Shillong, took up programmes for control of shifting cultivation which mainly aimed at soil and water conservation measures by forming terraces for development of agriculture and production of horticultural crops over an area of 11,360 ha at a cost of Rs 53.1 million (North-Eastern Council, 1978).

The status of shifting cultivation practice in the seven states of north-eastern region is given below (Anon, 1961–78).

Arunachal Pradesh

Shifting cultivation is mainly confined to the foothills and the middle belt in the entire state up to 2000 m elevation with an average rainfall of 2500 mm. Due to varying physiographic and climatic conditions, it is practised more in the eastern portion than in the western portion of the state. While the high-altitude dwellers like Monpas, Memba and Khamba do not practise shifting cultivation, the Sherdukpens of Western Kameng district depend partially on shifting cultivation. The Apatanis of Lower Subansiri have adopted an indigenous system of terrace cultivation near Ziro. All other major tribes, viz. Adi (including sub-groups of Gallong, Miniyong, Padam, Pasi, Bangni, etc.) Aka, Dafla or Nishi, Hill Miri, Khowa, Mishmi, Miji, Nocte, Tangsa and Wancho depend on shifting cultivation.

According to the 1971 census data, the tribal people constituted 78 per cent. Among Adis, the largest community of the shifting cultivators, the village council or Bango controls the distribution of hill slopes for shifting cultivation and the household enjoys the usufructary rights over the land. Though paddy is the main crop, millets, maize, yam, cotton, sugarcane, chilli, and tobacco are also grown. Shifting cultivation is practised for 1 or 2 years with a fallow period ranging from 3 to 10 years.

In the absence of adequate land records, the area under shifting cultivation is not correctly known. The average population density in the shifting cultivation area varies from 6 to 14 per km^2.

The shifting cultivation problem has been tackled in two ways in the state. Firstly, in the areas under shifting cultivation, the growing of low-volume, high-value cash crops like tea, coffee and big cardamom was encouraged in order to improve the economic conditions of the shifting cultivators. Secondly, the forest areas kept as reserve for future expansion of shifting cultivation were taken over by the Forest Department for scientific management under the Anchal Forest Reserve Scheme on equal sharing of the net profits with Anchal Samity. Under this scheme a residential school for 30 Mishmi girls has been started by the Tafragaon Anchal Samity, Lohit District, out of profits accrued during 3 years of management of 98 km^2 of Anchal Forests since 1976. More forest areas have been offered under this scheme, which is becoming popular. The North-Eastern Council programmes on soil conservation and horticulture were also effective.

Assam

Shifting cultivation is mainly confined to two districts, viz. Karbi-Anglong (Mikir Hills) and North Cachar Hills Autonomous Districts. While the North

Cachar Hills receive about 2500 mm. rainfall, the Mikir Hills, being in the rain-shadow region, receive 1200 mm rainfall. According to the 1971 census, the area has 25 persons per km^2.

The tribes practising shifting cultivation are Dimasa, Garo, Kachari, Karbi or Mikir, Kuki, Khasi, Lalung and Naga. While the Nagas and Kukis, being in hill areas, mainly depend on shifting cultivation, others depend on weaving, forest produce, silk-rearing, cotton-growing, etc. In the lower altitudes. Dimasas have taken up wet paddy cultivation also. Along with paddy, these people cultivate maize, vegetables, potato, cotton, ginger, chilli, yam, etc.

At village level, land is distributed by village council or village chief for shifting cultivation, and individual ownership is confined to permanent agricultural fields only.

In the hilly areas the shifting cultivation cycle is up to 10 years, while in the lower fertile areas it has become 2–3 years.

Under North-Eastern Council programmes, soil conservation and cash crop plantations like coffee were taken up to control shifting cultivation but the results obtained have been varied.

Manipur

The hills range from 800 to 3000 m elevation and have a rainfall varying from 1000 to 3000 mm per year. Shifting cultivation is practised in the five hill districts of the state. The population density in the area is 17 per km^2.

The tribal people are mainly Naga and Kuki-chin. They cultivate their plots for 1–2 years and then allow them to remain fallow for 6–10 years. Besides paddy, maize, potato, yam, chilli, vegetables, millets, mustard, til, sugarcane, ginger, turmeric and cotton are grown. In valley areas connected by roads, terraced cultivation has been taken up by the people.

Among some Kuki tribes, land is owned by the village chief who distributes the land to households on payment of annual tax. Otherwise land belongs to the community. Among the Nagas, land belongs to the village people and is distributed by village council, sometimes through lottery.

Through the North-Eastern Council programmes, soil conservation and cash crops like coffee and tassar silk cultivation on oak were taken up as alternatives to shifting cultivation, but the results obtained were variable.

Meghalaya

The state consists of the Garo Hills, the Khasi Hills and the Jaintia Hills in the form of a plateau. The annual rainfall varies from 12,000 mm at Cherrapunjee to 2500 mm at Garo Hill areas. The density of population is 45 per km^2. More than 80 per cent of the population is tribal and depends on agriculture.

Shifting cultivation is practised in the hill areas by the Garo, Khasi, Jaintia and Mikir tribal communities. Though people primarily depend on shifting cultivation, the intensity is varying.

The land is kept under cultivation for 2 years and left fallow for 4 – 5 years.

In the higher altitudes potato is the main crop, with paddy, maize, millet, yam, etc. as minor crops. In the lower altitudes paddy is the dominating crop, with potato, tapioca, yam, vegetables, oil seeds, millets, cotton, areca, betel, black pepper, banana, pineapple, citrus, etc.

The land belongs to the clan or to private individuals.

The use of soil conservation and cash crops like coffee, taken up in Garo Hills under North-Eastern Council as shifting cultivtion control programmes, was partly successful.

Mizoram

In this state the hills rise from 600 m in the west to 2300 m in the east, having an annual rainfall about 2500 mm. Almost the entire state is rural, with a population of 16 per km^2, and the economy is primarily based on shifting cultivation at subsistence level. This is one of the states with high literacy of 54 per cent.

Mizo, Paur, Lakhar and Hmar constitute the major Christian communities with Chakma and Riang belonging to other communities.

Under the Sixth Schedule of the Indian Constitution, the whole state is autonomous, managed by village councils. Every year the village council selects the hill area for shifting cultivation, divides it into plots and allots them by lottery to each household. Besides two crops of paddy, millet, sugarcane, chilli, ginger, turmeric, yam and tobacco are cultivated. The land is held under shifting cultivation for 1 to 2 years with a fallow period of 4 years. Terraced cultivation on the lower slopes and river valleys has been organized in places.

Through North-Eastern Council programmes, soil conservation and cash crop plantations like coffee were taken up in selected places for shifting cultivation control but they were not uniformly successful.

Nagaland

The hills of this state rise from 600 m in the west to 1800 m in the east, with an average annual rainfall of about 2500 mm.

The Naga tribe consists of different groups like Angami, Ao, Chakesang, Chang, Khiemungam, Lotha, Konyak, etc. Kuki and Dimasa are the two other important tribes.

The people mostly depend on agriculture based on shifting cultivation, though settled cultivation has also been developed to some extent. It has been estimated that on an average shifting cultivation is done for 2 years with a fallow period of 7–8 years.

The land is owned by village community or clan, or individuals. The rights of the chief over land are on the decline and the land is mainly owned by village communities. Besides the dominant crop of paddy, millet, potato, yam, chilli, maize and cotton are cultivated. The intensity of shifting cultivation is not uniform throughout the state, the percentage increasing from 10 per cent in the southern region to 20 per cent in the western region.

The cash-crop plantations like coffee and soil conservation programmes taken up under North-Eastern Council for control of shifting cultivation were generally effective and helpful.

Tripura

The state has low hill ranges rising from 230 m in the west to 900 m in the east, with an average rainfall of 2100 mm.

The tribal people occupy 29 per cent of the state. The density of population among shifting cultivators is 80 per km^2. Shifting cultivation is largely carried out in the western portion of the state by the tribal communities, viz. Riang, Chakma, Noatia, Jamatia, Tripuri and Lushai. The per capita area under shifting cultivation varies from 0.2 to 0.37 ha. After cultivating the land under shifting cultivation for 2 years, it is kept fallow for 7 years.

The revenue maps have been prepared by the state authorities. Besides long and short-term paddy, they also grow beans, pulses, cotton, jute, masta, cotton, oilseed and sugarcane.

The special scheme for settlement of shifting cultivators on integrated area development basis started by the Forest Department was found to be successful. Besides growing cash crops like rubber etc. they also grow fruit trees like banana, pineapple, citrus, jack, etc. The North-Eastern Council programmes on soil conservation and horticulture were also successful.

Land-use pattern

The experiments conducted by Indian Council of Agriculture Research, Shillong, regarding control of shifting cultivation indicated that in hilly areas, while the bottom one-third of the slope could be used for agricultural crops under terraced cultivation with irrigation facilities by having low-level dams and irrigation canals, in the middle one-third of the slope, horticulture crops like fruit trees, and cash crops like coffee and black pepper, could be grown along the contours and in the top one-third of the slope, forestry and fodder crops could be grown with pasture development along the contours for small timber, firewood and fodder. On this basis, a land-use pattern for shifting cultivation areas could be evolved.

In the development of a land-management system to replace shifting cultivation, the agro-ecological and socio-cultural factors in each area need proper emphasis, as there can be no one uniform pattern for all the areas. While the areas suitable for permanent cultivation should be developed with and without terraces for cereal, millet or cash crops with modern inputs, others may be suitable for forest plantation or maintenance as natural forests for fuel and wood production, or for agro-forestry.

POSSIBLE SOLUTION

Command area concept

The past experience has indicated that an integrated area development approach on watershed basis involving various disciplines directly connected with the people concerned—like agriculture, horticulture, forestry, animal husbandry, health, eduction and industry—would help us to provide an economically viable alternative to the people who practise shifting cultivation. Unless such an alternative could be provided to the shifting cultivators they would not be willing to give up the shifting cultivation by which they survive.

As far as possible, the areas selected should be viable units covering one ethnic group at least, as partial development of one ethnic group would generate imbalance within the group itself. Depending on the number of functional and non-functional activities, growth centres should be identified for development within the watershed command area. This would help us to utilize the available development fund properly to generate the maximum benefits to the people (Anon, 1961–68).

People participation

The shifting cultivators of the area should be associated in the formulation and implementation of the development programmes of their area from the beginning, by forming a consultative committee of local leaders, who could discuss and accept the programmes suitable to them. The object should be to secure the willing co-operation and involvement of the people to execute the programme. In finalizing the programme content, due weight should be given to the wishes of the people concerned.

Translocation of people to alternate areas has not only been costly but also not feasible in every case. *In situ* development has to be aimed at as the objective (Thangam, 1979).

Population control

In order that the maximum number should derive benefits out of the development programme within the minimum period, a family planning programme to control the population growth should be built into the programme itself. This has to be done on a persuasion basis without any compulsion, and this could possibly be best done by employing selected local people for training and executing it.

Cash crops

The development programme should aim at obtaining the best out of the land without detriment to the environmental factors of the locality. For this purpose, a land-use survey should be taken in the first instance to ascertain the best that could be grown in the land. In the development of land resources, the two important parameters—soil nutrients and water regime of the area—should be properly managed to get the best out of the land. The people should not only grow enough food crops by better inputs, but also improve their economic status. Such an economy could be developed not by growing cereals and millets but only by growing suitable low-volume, high-value cash crops like coffee, tea, cocoa, rubber, small and big cardamom, apples, spices like black pepper, cinnamon, medicinal and oil-bearing plants by adopting suitable agronomic practices.

The quick transformation brought about by adoption of cash crop economy can be visualized from what happened in Himachal Pradesh which was a deficit state in 1970. The apple gardens grown there helped the state to emerge as a prosperous rich hill state in a span of barely 6 years.

Multidisciplinary approach

The development programme should cater to the minimum needs of the people, viz. cereals, animal protein, fuel, fodder, clothing, housing, education, health and employment. In each discipline the programmes should aim at achieving the maximum result in the minimum period. In education, while children constituting 40 per cent of the population could be given formal education, the balance 60 per cent who form the workforce of the land should be educated through adult literacy programmes in their spare time. The indigenous system of medicine could also be made use of to provide adequate medical and health coverage including family planning facilities within the shortest period. In the social forestry programme, firewood plantations with short rotation should be grown to provide adequate quantities at a reasonable price to the people.

In order to provide maximum employment to the people, and thus improve their economic status, labour-intensive programmes, reducing mechanization in agricultural and other activities to the minimum, should be provided. The 'National Rural Employment Programme' could also be taken advantage of in this regard.

The traditional handicrafts like handloom cloth, woollen carpet, wood carving, metal works, cane and bamboo works, etc. should be developed through village, cottage and small-scale industries programmes, as an export market for such articles could be built up. Suitable local artisans should be selected and given training to develop these arts. They should also be encouraged to develop their own traditional arts, culture, dance and music in order to enable them to continue to have their colourful life.

Transition phase

The transition from the traditional custom-bound shifting cultivation to other modern methods of cultivation practices should be programmed in a smooth manner and at the pace of the shifting cultivator to have the minimum imbalance in their day-to-day living. They should be helped with abundant sympathy and kindness to adjust themselves to the new pattern with entirely different value-systems and pressures of development.

At times, a sense of post-contact demoralisation could set in among the people when the new system replaces their traditional age-old shifting cultivation and it can be as devastating as being deprived of the means of subsistence. Throughout the world, the indigenous minorities have lost much, if not all, that is positive in the dominant ones. Development often does not transpose a forest people into the mainstream of national society but into cultural limbo where it is neither one thing nor the other.

It is, therefore, necessary that the change-over should be at their own willing speed to help them to live 'with their own genius and tradition' (Poore, 1976).

National priority

An aspect which needs consideration is whether the shifting cultivator deserves more sympathy and attention than the urban slum-dweller or rural unemployed who live below the poverty line, when national priorities are concerned. Certain studies conducted in Maharashtra have shown that the shifting cultivator was more healthy than the urban or village poor. For over 9000 years shifting cultivation has been in existence and we have not yet been able to stop it completely. So, shifting cultivation practice has also to be taken as one of the problems which the nation has to solve by fixing due priority for it among the various national problems.

CONCLUSION

Shifting cultivation has remained a way of life among the tribal people and they would not willingly give it up unless a better alternative was shown to them. It was never a problem for the tribal people of the north-east region, or for all those who practised shifting cultivation all over the world. It has the forebodings of a problem to the outsider only. It would not be easy to provide the economically viable alternative of integrated area development programmes to all the tribal people practising shifting cultivation for a variety of reasons like lack of funds, defence needs, national priorities, etc. This human problem, though not insurmountable, has to be considered on merits in its own place among the priorities of the nation. However, in case of the north-eastern region it is a live issue which involves the major portion of its people and hence deserves topmost priority in the north-eastern regional development programmes.

REFERENCES

Anon, State Government Gazetteers and Census Reports—1961–78.

Bene, J. C., Beall, H. W. and Cots, A. (1977). *Trees, Food and People: Land Management in the Tropics*. International Development Research Centre, Canada.

Childe, V. G. (1956). *Man Makes Himself.* Watts & Co., London.

Conklin, H. C. (1961). 'A study of shifting cultivation. *Current Anthropology*, **2** (1).

Conway, Gordon and Romm, J. (1972) *Ecology and Resources Development in South-East Asia.* Ford Foundation, Bangkok.

North-Eastern Council Publications, Shillong, 1978.

Poore, M. E. D. (1976). 'The value of tropical moist forest ecosystems'. *Unasylva*, **28**, 112–13.

Thangam, E. S. (1979). 'Shifting cultivation in Arunachal Pradesh', *Proceedings of Agro-Forestry Seminar*, held at Imphal, India.

Socio-economic Effects and Constraints in Tropical Forest Management
Edited by E. G. Hallsworth

7

The effects of government policy in forest management in the Himalayan and Siwalik region of Uttar Pradesh, India

P. N. GUPTA
Uttar Pradesh Forest Department, Dehra Dun

INTRODUCTION

In the eight hill districts of Uttar Pradesh the population, as shown in the 1971 census, was 3,821,900. It was expected to rise to 4,629,000 by 1980, giving a population density of 75 persons per hectare. If the snow-covered areas of forest and the alpine pastures are excluded, since these are not inhabited for most of the year, the density of population by 1980 would be 250 to the hectare. This is high in relation to the low agricultural productivity of these areas.

The physiography of the region is varied. Altitudes range from below 300 to above 7500 m and the region contains the highest mountains and some of the biggest and most important rivers in India. The main Himalayan axis, running here south-east to north-west, comprises such major peaks as Nandadevi, Nandakot, Trisul, Kamet, etc. From this axis arise a number of major rivers which have their catchments separated by secondary mountain ranges. The rivers meander through the various mountain ranges through foothills to the Gangetic plain. The foothills merge into plains and flatten down into a narrow alluvial belt (known as Bhabar), running along their whole length and comprising boulder deposits of considerable depth with a very low watertable. Most of the minor streams dry up in the Bhabar tract, only to re-appear in a contiguous narrow long belt to its south, known as the Tarai. The Tarai has a shallow watertable, fine deep alluvial soil and luxuriant growth of coarse grasses. Some relatively wide and fertile valleys are found near the southern boundary of the tract, locally known as Duns. The Dehra Dun valley lies between the foothills and the Siwalik range, the latter being a range of low hills consisting of bouldery deposits to the south of the Himalayan ranges, formed by their erosion and alluvial deposition in the geological past.

About 93 per cent of the hill population lives in nearly 15,000 villages. Over 85 per cent of the rural people are cultivators and for every 100 men cultivators there are nearly 125 women cultivators. Cultivation is arduous and yields low return. Due to small holdings and their fragmentation, return from agriculture is further diminished. Most able-bodied men of the hills migrate to the plains for their livelihood, leaving behind women, children and old people to toil. To meet the family's requirements for food, firewood, fodder and water the women labour day and night in homes and at far-flung water sources, agricultural fields, pastures and forests situated on steep to precipitous slopes. The percentage of literacy in the hills is about 25.5, which is higher than the state average of 17.6.

The average per capita income during 1976–77 and 1977–78 in Uttar Pradesh was Rs. 809 and Rs. 892 respectively. For the hill region it is likely to be 60 – 75 per cent of that amount. This income accrues from various types of agricultural, horticultural, livestock, forestry and other commodities produced.

TIMBER, FIREWOOD AND GRAZING REQUIREMENTS

The local people's requirement of timber is for house construction, agricultural implements and packing cases. During 1977–78 the timber and firewood supplied free of cost to the people by the Forest Department was as shown in Table 1. With total projected population at 4.629 million, the number of families is nearly 1 million and their total timber and firewood requirement is estimated in Table 2.

Table 1 Timber and firewood supplied free

	Quantity (m^3)	Value (Rs.)	
Timber	85,772	15.06 million	16 per cent of total production
Firewood	208,442	0.22 million (= 1.18 million quintals)	

Table 2 Timber and firewood required

	Requirement per family	Total requirement
Timber	0.1 m^3 per annum	100,000 m^3
Firewood	10 kg per day or 3.65 quintals per year	3.65 million quintals

Table 3 Numbers of livestock

Type of livestock	Number (1974)	Cattle (unit per animal)	Total cattle units
Cows and oxen	1,447,546	1.0	1,447,546
Buffaloes	677,363	1.5	1,016,045
Goats	736,186	0.2	147,237
Sheep	353,198	0.2	70,640
Pones and mules, etc.	18,835	1.5	28,253
TOTALS	3,233,128		2,709,721

The number of livestock and their estimated requirement of fodder (for 8 hill districts) is as shown in Table 3. Allowing 70 per cent utilization factor annual fodder requirement at 4.5 kg per day per cattle unit

$$= \frac{2709721 \times 4.5 \times 365}{1000 \times 0.7}$$

$$= 6{,}358{,}166.7 \text{ metric tons, } or$$

$$= 6.4 \text{ million metric tons}$$

Agricultural crops produce at 1.28 metric tons of animal fodder per ha over nearly 480,000 ha of cultivated land or 0.60 million metric tons per annum. The balance requirement of animal fodder is thus estimated to be 5.8 million metric tons.

Assuming 2.5 metric tons/ha of production of fodder from present pastures, the total pasture land required for meeting the total fodder needs comes to 2.32 million ha (23,200 km^2). This is nearly 50 per cent of the total area of the hill portion. Much of the land is neither under pastures nor can it be brought under pastures. The result is that the existing pastures and forests are overgrazed and the cattle and other stock underfed.

PEASANTS RIGHTS AND CONCESSIONS

The story of rights and concessions is old and complex and even the British rulers accepted the position as too difficult to change. The existing rights make it extremely difficult to manage the forests efficiently. Almost all the villages have certain prescribed rights and concessions as recorded since Traill demarcated the boundaries of land and villages for revenue administration during 1817–23. These vary in different classes of forests

Old Reserve (1000 km^2) Declared as such in 1877 near Nainital, Almora and Ranikhet. Less rights in respect of grazing, grass-cutting and leaf-plucking.

Class I New Reserves Constituted during 1911–17 but classed as such on the recommendations of Kumaon Grievance Committee of 1921. *Placed under Revenue Administration*. Right-holding village people can cut any tree of any species and lop branches of the same, except the protected trees (chir, sal, deodar, surai, kail and spruce), for which permission of the District Magistrate was required.

Class II New Reserves Right-holding residents of Kumaon can under certain restrictions cut trees other than protected. These were placed under the Forest Department.

People have always been agitating for relaxing of restrictions and wanting extension of their rights and concessions. As a sequel to recommendations of the Kumaon Forests Committee (1932), rules to regulate these were framed in 1941 and *fire protection was altogether withdrawn from Class I forests* where grazing, felling of trees and lopping was also allowed without limit or restrictions. These forests were supposed to have little commercial value for timber, but were of special importance as containing the sources of water supply to the chief rivers. In Class II forests, which were to be protected from fire, there was little difference in management except that uncontrolled felling of the above-mentioned protected species was restricted. Indents for these were to be submitted to the Forest Department and would be supplied to the extent of recognized rights.

Even these concessions, allowed from time to time, did not satisfy the people, and in 1937 the then Chief Minister, Pandit G. B. Pant, wrote:

> We hold that, where a considerable time had passed since the reservation of forests, it is proper to revise the rights and to see if, consistently with the interests and demands of forests, they can stand a larger and heavier strain, and in case they can, to enlarge the rights that may have been allowed previously.

The administration problem is that it is the common belief that forests are gifts of Nature for the unrestricted use of the people. Any restrictions imposed for the protection of the forests are considered to be a forfeiture of their natural rights. For looking into and investigating the grievances of the inhabitants of this region, a Kumaon Forest Fact Finding Committee was appointed in 1959 and it went into this whole question in great detail.

THE LEGAL STATUS OF THE FORESTS

The legal status of the forests for 8 hill districts is approximately as shown in Table 4. Out of the forests that are under the control of the Forest Department, 27 km^2 are protected forests, 110 km^2 unclassed forests and the rest are reserved forests. The reserved forests were previously classified into old and new reserves. Then later, they were again classified into Class I and Class II

Table 4 Areas of each category of forest

	km^2
(1) Area under the control of Forest Department	23,766*
(2) Civil and Soyam forests	7,607
(3) Panchayati forests	2,448
(4) Private, municipal, etc.	232
TOTAL	34,053*

*This figure is higher than given earlier because the legal forests include denuded areas, snow, and also alpine pastures.

about 1921 as a result of public agitation against the hardships of the people resulting from the then forest administration. This classification was in existence in the old Kumaon Forest Circle, now split into Kumaon and Garhwal Circles. Class II reserved forests and old reserves comprise the more commercially valuable forest with good protection and supervision of the Forest Department, whereas Class I reserves comprise forests then considered of less commercial value, which were placed under the Revenue Authorities. In Class II reserves, at any time more than one-sixth of the total area of any reserve cannot be fenced for plantations, afforestation or natural regeneration without special sanction of the state government. Civil forests legally are protected forests. The Panchayat forests include both reserved and protected forests. Their formation has to receive sanction of the Divisional Commissioner.

Forests other than reserved

To further meet the requirements and needs of the local people, there were protected forests, known as civil forests in Kumaon, and these were managed by the Revenue Authorities for the benefit of the people. Similar forests in erstwhile Tehri Garhwal State, which merged with Uttar Pradesh in 1948, are known as Soyam forests. At one time the condition of these forests was also like that of the reserved forests but most of these forests have been destroyed by uncontrolled exploitation by the village people.

Panchayat forests

The Kumaon Grievances Committee in 1921 had recommended that the forests left for villages should be placed under the Panchayats. In 1930 Forest Panchayat Rules were made and forest panchayats were formed for Civil and Class I forests. The condition is generally satisfactory and the income can be spent by the people on works of public utility.

THE IMPACT OF GRAZING

An annual migration of population is a special feature of this tract. During winters, poorer residents of middle and higher hills migrate to Bhabar, along with their cattle and entire familes. In Bhabar they earn wages as agricultural and forest labour and return to their hill homes during the spring. Similarly Bhotias and Jads, residents of inner hills, also migrate along with their sheep and goats to lower hills, Tarai and Bhabar. Apart from these a number of Jammuwala gujars have been living in the tract for a number of years as migratory graziers. They keep buffaloes for milk. They migrate to sub-alpine pastures during summers and return to Siwaliks or Bhabar during winters.

Local graziers

These are mostly residents in villages and have rights to graze their cattle in the forests. In the Tarai, Bhabar and plains portion the number of cattle allowed to graze free is fixed at the time of forest settlement, and the excess number is charged nominal grazing fees. In the hills there is no restriction as to number or locality where grazing can be practised. The hill villagers have rights to make temporary cattle stations (Kharaks) in forest areas in certain seasons of the year annually.

Migratory graziers

There are several categories

(1) Jammuwala Gujars: there are nearly 400 families with a total population of 3500 – 4000 and the number of buffaloes with them is about 15,000.
(2) Bhotias: their total population is 10,620 and an average family possesses 100 – 150 sheep.
(3) Gadis: migratory shepherds of Himachal Pradesh who come to Dehradun district. Jads are similar people of Uttarkashi District. The total number of their combined sheep is estimated to be about 20,000.
(4) Ghamtapas and Goth-walas: other graziers who rear their livestock within the forests.

FOREST MANAGEMENT

Early history

According to the available records, planned forest management is said to have started only after the advent of British administration. In the beginning the population was sparse and forest area extensive. The local people could satisfy all their needs from the forests. The so-called scientific forest management has been confined to the area under the control of the Forest Department. Even within this area a category of forest known as Class I forests came under the charge of the Forest Department sometime at the end of 1964 when the Government observed that

the deterioration which had set in the condition of Class I forests since their transfer to the management of the Revenue Authorities had steadily aggravated and these forests (in 1964) were in a state of almost complete extinction. This had adversely affected the economy of Kumaon and [the Government felt that] if allowed to continue, would have irreparable consequences on the economic conditions of the Kumaonies.

Kumaon, Tehri State and other forests have had a different early history which in brief is as follows.

Kumaon forests

British administration started in 1815. During 1828–58 chir, even though in abundance, was not considered of any commercial importance as sal was available from areas closer to the plains. Sal was arbitrarily felled for railway sleepers. Sir Henry Ramsay, Commissioner for Kumaon, was asked to manage forests (1858–68), and a separate Conservator was appointed in 1873 when the Reservation and Protection of forests was started; this was completed by about 1915. By 1925 written Working Plans had been prepared for all the reserved forests. According to Canning the basic principle of the exploitation was to consider the interest of the purchasers of forest produce and to see that they made good profits, and exploitation of forests proceeded rapidly. The great increase in pine timber used resulted in large departmental open-tank creosoting intallations designed to treat over a million railway sleepers in 3 years. This was before the First World War. Resin tapping had been in progress for some years in old reserves near Bhowali and Nainital when it was started in Almora district in 1911. By the end of the First World War the resin outturn from Almora had risen to 35,000 quintals. Control-burning as a fire conservancy measure was introduced at about the same time. The earliest scheme was for chir fellings in Kali Valley (1912–15).

Dun and Siwalik forests

Because of indiscriminate fellings, Wallich pointed out the need for conserving the forests of sal and shisham at the Himalayan foot hills. However, as early as 1825, Falconer's observation that large numbers of seedlings occurred every season near the big trees led him to conclude that the forest wealth was inexhaustible. The difficulty of raising sal was not realized then, and this belief was only dispelled after more and more exploitation.

The initial interest in collection of revenue only meant hardly any control on fellings. In 1878 O'Callaghan recorded for Dehradun forests that 'there can be no doubt that sal, tun and sissoo were the trees chiefly felled, for even now there is no demand for any other kind of timber and when I entered the department in 1864 the ground was studded with stumps of these trees everywhere'. The system followed was absolutely devoid of conservancy and consisted of colossal destruction and waste of forest resources which abounded in fine trees 100 – 200 years old and upwards. Brandis' guideline for the Siwalik division came in 1877 while the first working plan of Dehradun Division was that of Fernandes (1888).

Tehri – Garhwal forests

One Mr Wilson had lease of exploiting forest produce from the then Maharaja (latter part of the first half of the last century to 1850 and then from 1850 to 1965, leased for Rs. 400 per annum) and valuable deodar forests were arbitrarily felled and destroyed. Mr Wilson was the first man to start river transport, and started making railway sleepers. The Government of North Western Province (the present Uttar Pradesh) took the subsequent lease for Rs 10,000 per annum and Tons, Yamuna valley forests on Rs 4500 per annum. At the expiry of this lease in 1885, the state established its own Forest Department. Survey and boundary demarcation started in 1880. In 1897 the State requisitioned the services of two trained officers and management was based on forest preservation.

Boundaries of valuable forests were demarcated and restriction imposed on the use of forests by village people. Rules for felling, lopping and fire protection were made, and the previously unrestricted exploitation of forests was brought to an end. In 1905 these rules were relaxed due to opposition by villagers. Forest settlement operations were carried out from 1909 to 1911 and 1929 to 1931, and forests were divided into reserved and open (*soyam*). The State forests merged with Uttar Pradesh in 1948 and the condition of *soyam* forests is as bad as that of civil forests.

Socio-economic Effects and Constraints in Tropical Forest Management
Edited by E. G. Hallsworth

8

Socio-economic aspects of forest resources management in Indonesia

ISHEMAT SOERIANEGARA
Director, SEAMEO Regional Center for Tropical Biology (BIOTROP), Bogor, Indonesia

INTRODUCTION

To fulfil its role in solving today's problems forest management should not only deal with the economic and financial aspects of forestry. It should be expanded to become forest resources management (Soerianegara, 1978; Duerr *et al.*, 1979).

In this new concept forest resources management is to be regarded as a segment of the natural resources management, the object of which is the natural resources ecosystem, in which human populations constitute active components in the production, the distribution, the consumption and the development of the ecosystems and resources. The relationships between the natural ecosystems/resources and the human populations is depicted in Figure 1. Obviously, the wise use and sound management of the forest resources aim at the balance between the utilization and the development of resources. The success in reaching this balance depends on the economic and environmental policies, the application of appropriate technologies and rational economic and ecologic consideration, the presence of skilled manpower and the socio-economic environment, including the correct attitude of the public. This is illustrated in Figure 2.

FOREST RESOURCES IN INDONESIA

Forest is the major natural vegetation in Indonesia and is a very important natural resource. The extent of the natural forest is around 106.5 million ha (Directorate of Forestry Planning, 1979), consisting of six major forest for-

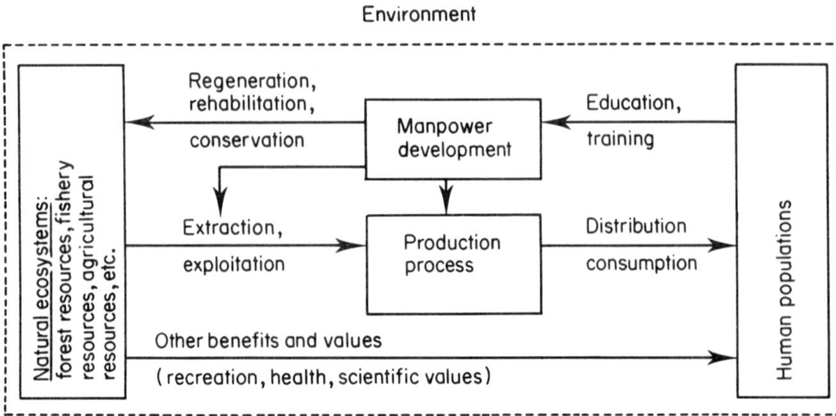

Figure 1 The natural resources ecosystem

mations, namely: beach forest, mangrove forest, swamp forest, peat-swamp forest, rainforest and monsoon forest. The extent and distribution of these forests are shown in Table 1.

In addition, there are around 13.9 million ha of secondary forests and around 1.8 million ha of forest plantations (teak, pine, *Agathis, Altingia*), so that the total forest area is around 122.2 million ha. Compared to other kinds of land-use, forest land occupies the largest land area, namely 63.7 per cent of

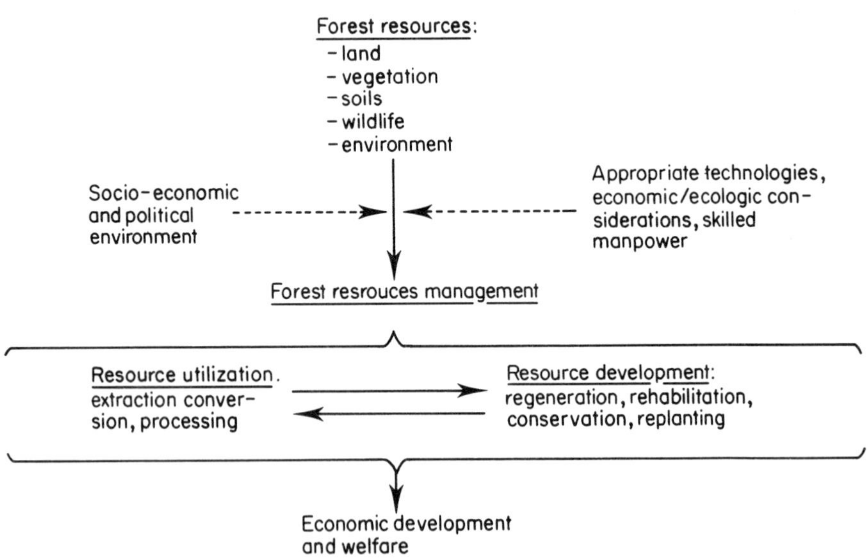

Figure 2 The goals and constraints in forest resources management

Table 1 The extent and distribution of the major forest formations in Indonesia

Forest formation	Extent (million ha)	Major areas of distribution
Beach forest	1.0	Sumatra, Java, Bali, Sulawesi
Mangrove forest	1.0	Sumatra, Java, Kalimantan, Irian Jaya
Swamp forest	13.0	Sumatra, Kalimantan, Irian Jaya
Peat-swamp forest	1.5	Sumatra Kalimantan
Rain forest	89.0	Sumatra, Kalimantan, Irian Jaya
Monsoon (seasonal) forest	1.0	Java, Nusa Tenggara, Sulawesi

the total land area (see Table 2). The extent of forest land on each island of the Indonesian archipelago are shown in Table 3. The forests in Indonesia are also classified according to their functions into: protection forest, production forest, conservation forests (nature conservation forest and recreation forest) and reserved forest. The extents of these forests are given in Table 4. To get

Table 2 Land-use

Kinds of land use	Extent (million ha)	Percentage of land area	Ha per capita*
Forest land	122.2	63.7	0.85
Agriculture	13.9	7.2	0.09
Estate crops	1.5	0.8	0.01
Other uses	54.3	28.3	0.37
TOTAL LAND AREA	191.9	100.0	1.33

*Based on the estimated population of 144.2 million.

Table 3 The distribution of forest lands in the Indonesian archipelago (Directorate of Forestry Planning, 1979)

Island(s)	Land area (million ha)	Forest land (million ha)	Forest land (percentage of land area)
Sumatra	47.4	28.4	59.9
Java-Madura	13.2	2.9	21.9
Kalimantan	53.9	41.5	76.9
Bali and Nusa Tenggara	8.8	2.0	22.7
Sulawesi	18.9	9.9	52.4
Moluccas	7.5	6.0	80.0
Irian Jaya	42.2	31.5	74.6
TOTAL	191.9	122.2	63.7

Table 4 Forests classified according to their functions (Directorate of Forestry Planning, 1979)

Kinds of forest	Planned		Existing	
	Area (million ha)	Percentage from total forest area	Area (million ha)	Percentage from total forest area
Protection forest	47	38.5	16.7	13.6
Production forest	40	32.8	59.2	48.5
Conservation forest	10	8.2	6.6	5.4
Reserved forest	25	20.5	39.7	32.5
TOTAL	122	100.0	122.2	100.0

an idea about the production potential of the forests, Table 5 summarizes the production potential of the forest plantations and the natural mixed forests in Indonesia. More detailed data on the production potential of the natural forests are given in Appendices I and II at the end of this chapter.

FOREST RESOURCE UTILIZATION AND ITS EFFECTS

Logging is the foremost method of forest resource utilization. According to article 33 of the Indonesian Constitution, natural resources, such as forest resources, should be wisely used for the welfare of the people. Forest resources utilization is regulated by Law No. 5/1967 concerning the Fundamental Regulation on Forestry. The Indonesian government has given opportunity to both domestic and foreign capital investors to utilize and develop natural resources in Indonesia. It is regulated by Law No. 1/1967 concerning Foreign Capital Investment and Law No. 6/1968 concerning Domestic Capital Investment.

The government of Indonesia has also issued Government Regulation No. 33/1970 concerning Forest Planning, and Government Regulation No. 21/1970 concerning Forest Utilization Rights and Forest Products Exploitation Rights. In addition, forest utilization conducted by private companies should follow the stipulations in the Forestry Agreement between the investors and the government, which is represented by the Director General of Forestry.

Logging operations are generally conducted by private companies according to the approved Forest Operating Plans, consisting of the Twenty-Year Plan, the Five-Year Working Plans and the Annual Working Plans. Logging techniques and forest regeneneration methods are stipulated in the Letter of Decision of the Director General of Forestry No. 35/1972. Data in Appendices I and II indicate that the total standing stock of commercial timber species in natural forests is estimated at 3600 million m^3 from about 37 million ha of natural forests with an average volume of 97.4 m^3 per ha.

In Table 5 it is estimated that the production potential of the production forests (natural and plantation forests) is around 2600 million m^3 from about 30 million ha of forests. If the expected log production is roughly 40 per cent of the potential production, and the cutting cycle is 35 years, the expected log production will be around 1050 million m^3 logs, with an annual production of about 30 million m^3. The actual timber production details during 1972–78 are shown in Table 6. Although the potential or expected annual production was not exceeded by the annual timber production, in some areas overcutting has taken place, which endangered the continuous production of certain timber species. Soedjarwo (1978) summarized the following positive and negative effects of logging:

(1) The positive effects are:

- (a) Increase of the GDP of forestry amounting to 10–26 per cent, and the national GDP by 7 per cent.
- (b) Increase of the State revenue of foreign exchange, which places forestry into the second position after oil in terms of foreign exchange earning (US $750 million in 1974).
- (c) Expansion of job and business opportunities. In 1975 there were already 240 forest concession holders which absorbed a labour force of around 24,000 people. If all the 630 forest concessions are in operation, they could absorb about 63,000 people.
- (d) Increase of the local people's income, due to the creation of new agricultural and business opportunities.
- (e) The establishment of communication facilities by the logging companies increases the economic mobility of the local population.
- (f) Increase of skill of Indonesian personnel.
- (g) Acceleration of social growth of the local community, due to the establishment of public health centres, schools, mosques, etc.
- (h) Funds become available for forest inventory and replanting and also for resettlement schemes for the shifting cultivators.

(2) The negative effects are:

- (a) Since logging is now restricted to certain timber species, this may result in over-cutting of these species and cause degradation of the forest composition and productivity.
- (b) Wages paid in logging companies are relatively high, so that this may disturb other economic activities.
- (c) Log skidding, which frequently violates existing regulation, would result in the destruction of the residual stands and cause soil erosion.

Table 5 Production potential of the productive forests (forest plantation and natural forests) for 1972–91 (Directorate General of Forestry, 1973)

Islands	Forest plantations							
	Teak		Conifers		Other hardwoods		Total	
	Area (1000 ha)	Volume (1000 m^3)	Area (1000 ha)	Volume (1000 m^3)	Area (1000 ha)	Volume (1000 m^3)	Area (1000 ha)	Volume (1000 m^3)
Sumatra	—	—	150	15,000	24	2,400	174	17,400
Java	761	76,100	145	14,500	85	8,500	991	99,100
Kalimantan	—	—	—	—	—	—	—	—
Sulawesi	7	525	—	—	4	300	11	825
Mollucas	—	—	—	—	—	—	—	—
Irian Jaya	—	—	—	—	—	—	—	—
Bali and Nusa Tenggara	6	300	3	225	18	1,170	27	1,695
INDONESIA	774	76,925	298	29,725	131	12,370	1,203	119,020

Table 5 *(continued)*

Islands	Natural forests					Grand total	
	Commercial tree species		Other species	Total			
	Exportable	Domestic					
	Volume (1000 m³)	Volume (1000 m³)	Volume (1000 m³)	Area (1000 ha)	Volume (1000 m³)	Area (1000 ha)	Volume (1000 m³)
Sumatra	258,510	324,984	59,008	7,386	642,582	7,560	659,982
Java	—	—	—	—	—	991	99,100
Kalimantan	611,640	428,156	198,783	15,291	1,238,571	15,291	1,238,571
Sulawesi	38,961	28,971	5,994	999	73,926	1,010	74,751
Mollucas	140,940	7,830	15,660	1,566	164,430	1,566	104,430
Irian Jaya	91,600	32,976	285,792	3,644	403,430	3,664	403,430
Bali and Nusa Tenggara	378	1,358	196	14	1,932	41	3,627
INDONESIA	1,142,029	824,205	565,433	28,920	2,524,871	30,123	2,643,891

Table 6 Timber production, 1972–78, as equivalent m^3 roundwood (Directorate of Forestry Planning, 1979)

Year	Roundwood	Sawnwood	Total
1972	16,877,000	840,000	17,717,000
1973	24,919,950	1,377,345	26,297,295
1974	20,860,961	892,060	21,753,021
1975	14,587,558	1,708,015	16,295,873
1976	20,803,052	624,841	21,427,893
1977	22,334,844	604,803	22,038,848
1978	24,742,900	1,512,742	26,255,643

(d) The opening up of a forest road network may encourage shifting cultivation.
(e) Logging activities change the forest environment and may affect the life of wild animals.
(f) Without enforcing the existing regulations, logging will result in the destruction of the residual stands, the radical change of the tree species composition and leave an abundance of waste in the logging area.

The logging effects could also be classified into ecological and socio-economic effects.

Ecological effects

Logging causes the decrease in biological diversity, stability and productivity; the decrease in soil fertility, water quantity and quality; the increase of soil erosion and surface run-off, and the decrease in the general quality of the environment.

It is an established fact that certain valuable tree species like *Manilkara kauki, Diospyros celebica* and *Gonystylus bancanus* are becoming scarce in Indonesia. The natural population of extensively planted *Agathis loranthifolia* has been lost. Extensive areas of secondary forests, savanna-like vegetations and *Imperata* (alang-alang) grasslands have developed following logging, shifting cultivation and burning.

Logging methods certainly influence the degree of damage. Haphazard logging will produce the most harmful effect. In many cases cable yarding is more destructive than tractor yarding. However, whole-tree logging by tractor will be most harmful.

Socio-economic effects

Logging and forest product industries, which produce goods beneficial for human welfare, could produce unfavourable environmental impact. A balance should be established between socio-economic benefits and ecological quality.

In many cases, however, unfavourable ecological effects are accompanied by unfavourable socio-economic effects. Loss in resource productivity, decrease in environmental quality, and in aesthetic as well as scientific values are unfavourable ecologically and socio-economically.

A study on the socio-economic impact of logging in Riau (Wirapradja *et al*., 1970) reveals that logging caused the transmigration of labourers from outside Riau, the increase of income of local population and the increase of revenue of the provincial government. However, the loss inflicted by the ecological impact caused by uncontrolled logging might exceed the beneficial socio-economic impact. Thus, a study of logging effects should comprise ecological and socio-economic aspects.

PROBLEMS IN FOREST RESOURCES MANAGEMENT

From the previous discussion it is obvious that in connection with logging operations the following problems arise:

(1) enforcement of forestry regulations concerning logging;
(2) control and monitoring of logging activities;
(3) forest regeneration, tending, replanting;
(4) recruitment of foresters, forest rangers and skilled workers.

The shortage of skilled manpower to support forest resources utilization and management is serious. At present it is estimated that the ratio between university graduate foresters and the forest area is about 1 : 300,000 ha.

The problems are not only concerned with logging and with forest lands. Different kinds of problems exist in Java and in other areas outside Java. In Java, the most densely populated island, population exerts a heavy pressure on forest land. Here the following problems require attention:

(1) establishment, management and protection of forest;
(2) watershed management to prevent floods and soil erosion;
(3) reforestation of forest land and revegetation of unproductive agricultural lands;
(4) rehabilitation of critical lands;
(5) provision of fuelwood.

Arising from the inadequate supplies and control, the illegal cutting for timbers and fuelwood from forest plantation, protection forest and conservation forest constitutes a major problem. In areas outside Java the following problems prevail:

(1) regeneration, replanting and tending of logged forest areas;
(2) shifting cultivation and resettlement schemes;
(3) rehabilitation and conversion of alang-alang grasslands and waste lands;
(4) establishment and management of conservation forest;
(5) conversion of natural forests into transmigration settlement areas and new agricultural areas.

SOCIO-ECONOMIC ASPECTS

Although there are several problems which influence forest resources management, for the purpose of this chapter I will concentrate on the socio-economic aspects.

The general socio-economic conditions which have some influence on forest destruction are the general poverty and lack of proper education of the local population.

In Java illegal cutting for timber or for fuelwood is done by the local people to earn some money or to fulfil their needs; and these people do not understand the ecological consequences of forest destruction. In reforestation, the local people who are employed in the taungya (tumpang sari) system, tend to damage the young trees in order to prolong their stay on the forest land to raise their crops. In revegetation of unproductive agricultural lands, the local people also tend to damage the trees because of their need for fuelwood, they dislike the tree species or they are against the rehabilitation project because they could not grow their own crops. To minimize these problems people from very densely populated and poor areas in Java have been transmigrated to the islands outside Java, especially Sumatra and Kalimantan.

In the islands outside Java shifting cultivation and forest burning, which have created extensive alang-alang and shrub waste lands, are still going on. The forest road network created by the logging companies has promoted the spread of shifting cultivation. Therefore the rate of reforestation and rehabilitation of waste lands cannot keep pace with the rate of forest destruction caused by logging activities and shifting cultivation.

Another cause of serious forest destruction is the opening up of natural forests for transmigration projects, which, in general, has not been properly planned and executed. Some of these areas which were natural forests have been wasted.

Upland forest and vegetation destruction has caused floods and soil erosion which endangered agriculteral lands, village settlements and dams or water reservoirs.

All too often the forest service (the Directorate General of Forestry) has been blamed for not managing the upland forests properly. But the bare lands in the uplands, especially in Java, do not belong to the Forest Department; instead they belong to the local communities or are privately owned.

Therefore, forest destruction and its effects do not and should not only concern the forest service but also other government agencies and ministries. Integrated efforts should be conducted to solve the problems.

At present the Indonesian government has launched two big programmes under the direction of the Minister of State for the Supervision of Development and the Environment (1979):

(1) the Forest, Land and Water Conservation Programme;
(2) the Natural Resources and Environment Development Programme;

involving several departments (ministries) and agencies. With regard to the shifting cultivation in the islands outside Java the Indonesian government has launched the Resettlement Project, also involving several departments. As any other development projects those programmes and projects will again have socio-economic impacts.

At present very few researches have been conducted in the socio-economic aspects of forest resources management. The socio-economic researches should explore at least the following:

(1) the values of the forest resources for the local community;
(2) the attitude of the local community towards the original forest resources and the development activities (reforestation, revegetation, rehabilitation and resettlement);
(3) the benefits of development activities for the local communities;
(4) local community participation in the development of forest resources (including nature and wildlife conservation).

Appendix I Average volume per ha of important timber trees in natural forests, and their maximum annual allowable cut (MAAC) for each province, based on surveys conducted during 1966–74

Province	Total forest lands (1000 ha)	Forested lands (1000 ha)	Average volumes, m^3 per ha (trees of 50 cm diameter and over)						(MAAC) (1000 m^3)
			Dipteroc	Non-Dipterocarp	*Ramin*	*Agathis*	Others	Total	
Aceh	1,284.90	1,067	89.29	12.21	—	—	15.59	117.09	2,145
N. Sumatra	850.75	539.25	54.23	4.14	—	—	21.49	79.86	1,251
Riau	5,983	4,668	30.75	24.90	—	—	17.17	72.82	3,502
W. Sumatra	878	623	66.40	21.66	—	—	16.62	104.68	904
Jambi	3,951	2,709.50	37.03	33.16	0.51	—	10.97	81.67	2,667
S. Sumatra	2,577	1,734.50	26.49	33.19	0.28	—	8.60	68.56	1,324
Bengkulu	623	395	62.17	29.54	—	—	25.16	116.87	689
Lampung	310	214	30.15	20.83	—	—	25.93	76.91	242
W. Kalimantan	4,532.50	2,549.65	41.57	24.81	8.77	0.04	16.29	91.48	2,554.50
C. Kalimantan	10,780	8,344.25	43.82	11.66	5.67	0.15	5.70	67.00	9,178.50
S. Kalimantan	818	489.25	42.67	27.96	1.61	—	11.53	83.76	574.50
E. Kalimantan	10,522.50	8,308.70	79.72	40.29	0.01	0.52	25.18	145.72	14,136
N. Sulawesi	620	552	1.94	56.58	—	—	27.29	85.83	433
C. Sulawesi	1,929	1,478	0.55	55.83	—	0.72	35.17	95.27	1,260.50
S. Sulawesi	708	561.50	3.74	74.47	—	1.00	22.57	101.78	835
S.E. Sulawesi	438	376	0.13	35.17	—	2.14	18.61	56.05	232
W. Nusa Tenggara	87	25	—	4.36	—	—	2.91	7.27	87
E. Nusa Tenggara	—	—	—	—	—	—	—	—	—
Moluccas	2,644	1,875.65	60.59	32.14	—	0.78	20.04	113.55	2,579
Irian Jaya	998.50	891	0.40	31.51	—	0.27	29.01	61.19	552
TOTAL	50,458.15	37,401.25	47.64	29.13	2.94	0.40	17.25	97.36	45,164

Source: Central Bureau of Statistics, 1975 (Statistics on Indonesia); Jakarta.

Appendix II Estimated standing stock of important timber species in natural forests, based on preliminary surveys conducted in 1966–74, thousand m^3 (trees of 50 cm diameter and over)

Province	Dipterocarpaceae	Non-Dipterocarpaceae	*Ramin*	*Agathis*	Others	Total standing stock
Aceh	95,272	13,028	—	—	16,634	124,934
N. Sumatra	29,244	2,233	—	—	11,589	43,064
W. Sumatra	41,367	13,494	—	—	10,354	65,215
Riau	143,541	6,233	—	—	80,150	339,924
Jambi	100,333	89,874	1,382	—	29,723	221,311
S. Sumatra	45,949	57,568	486	—	14,919	118,902
Bengkulu	24,557	11,668	—	—	9,982	45,607
Lampung	6,152	4,577	—	—	5,549	16,578
W. Kalimantan	105,989	73,455	22,360	102	41,534	243,440
C. Kalimantan	365,645	97,940	47,312	1,252	47,562	559,711
E. Kalimantan	662,370	334,758	83	4,321	209,213	1,210,745
S. Kalimantan	20,878	13,675	788	—	5,641	40,982
W. Nusa Tenggara	—	109	—	—	73	182
N. Sulawesi	1,071	31,232	—	41	15,064	47,378
C. Sulawesi	813	86,951	—	1,064	51,981	140,809
S. Sulawesi	2,100	41,815	—	562	12,673	57,150
S.E. Sulawesi	49	13,224	—	805	6,997	21,075
Moluccas	113,646	60,283	—	1,463	37,588	212,980
Irian Jaya	358	28,075	—	241	25,480	54,154
TOTAL	1,759,633	1,090,191	72,411	9,821	632,106	3,564,160

Source: Central Bureau of Statistics, 1975 (Statistics on Indonesia); Jakarta.

REFERENCES

Directorate of Forest Planning (1979). *Forestry in Indonesia 1978*. Bogor, Indonesia.

Directorate General of Forestry (1973). *Draft Forestry Master Plan 1972–1991*. Jakarta, Indonesia.

Duerr, W. A., Teeguarden, D. E., Christiansen, N. B., and Guttenberg, S. (1979). *Forest Resource Management*. W. B. Saunders, Philadelphia.

Minister of State for the Supervision of Development and the Environment (1979). *Policy and Programs – Management of Natural Resources and Environment in Repelita III*. Jakarta, Indonesia.

Soedjarwo (1978). Keynote address. *Proceedings Symposium on the Long-term Effects of Logging in South-East Asia*, pp. 11–17. BIOTROP, Bogor, Indonesia.

Soerianegara, I. (1978). *Forest management and conservation, or forest resources management*. Special Paper, 8th World Forestry Congress, Jakarta. Directorate General of Forestry, Jakarta, Indonesia.

Wirapradja, A., Hadikoesworo, H., and Soemarmo, S. (1970). *The Impact of Forest Exploitation on the Socio-economic Conditions in Riau*. Report No. 115. Forest Products Research Institute, Bogor, Indonesia.

Socio-economic Effects and Constraints in Tropical Forest Management
Edited by E. G. Hallsworth

9

The socio-economic effects of forest management on the lives of people living in the area: the case of Central America and some Caribbean countries

GERARDO BUDOWSKI
CATIE, Turrialba, Costa Rica

INTRODUCTION

In Central America and the Caribbean and for the purpose of this account, the various types of forest management can roughly be divided into four categories according to land use:

Protection of the forest

There are two broad sub-divisions:

(1) Total protection. Aimed at and achieved in national parks, certain critical catchment areas and analogous well-protected areas. Preservation of valuable ecosystems and protection of water flow are the main objectives. Except for some national parks, very few other areas are presently well protected.
(2) Partial protection. Usually when the responsible authority is not in position really to control land-use, even if the need (and the declared policy) calls for total protection. A large amount of land falls under this category.

Managing forest land for protection is often considered a restriction to free land-use by many local inhabitants, although there are benefits to people, especially those depending on the water outside the locality. With increasing awareness of soil erosion there is at present considerable interest in this type of management but also some confusion as to how it should operate.

When other land-uses are considered compatible with the protection, they will be examined under multiple use or in some cases under agroforestry.

Forest management where wood production is the main object

Two main divisions can be considered: production from natural forests (including primary forests, secondary forests, swamp forests etc.) and from plantations.

(1) Natural forests. A very large amount of forest products in Central America and the Caribbean come from natural forests and the impact on people is considerable.
(2) Plantations. Only small areas are in forest plantation, possibly not even 15,000 ha in the whole of Central America and the Caribbean. Agroforestry combinations which cover considerably larger areas are not discussed here but are a separate category. The impact of plantations on the lives of people is relatively small but it is likely to increase considerably in years to come.

Agroforestry

Some confusion exists in terminology and many definitions are found (see ICRAF, 1979, p. 4). As used for years at CATIE, the term agroforestry implies land management techniques involving the combination of forest trees with crops, or with domestic animals, or both. The combination may be either simultaneous or sequential in time or in space. The aim is to optimize production per unit area while at the same time respecting the principle of sustained yield (Combe and Budowski, 1979).

Trees are usually planted, or at least tended, for their products or their 'services' (the latter implies soil improvement or restoration, protection against wind, erosion, excessive sunlight, etc.). Included here are large areas where coffee and cocoa are cultivated under shade trees (that also provide firewood and other products) and other combinations of trees with annual or perennial crops, as well as combinations of trees with pastures or cut-grass, trees for fodder, live fenceposts, windbreaks and other uses of trees for protective purposes.

Agroforestry has a very large impact on the lives of people in the region and will be examined in some detail because of its possibilities of improving the conditions of rural communities.

Multiple use

Included here are combined uses not considered previously, where besides forest management for wood, management also involves protection of watersheds, provision of recreation facilities in natural forests or plantations, grazing in natural forests, or hunting and fishing in forest land.

The impact is spotty. The most important case of multiple use probably is found in east Puerto Rico (Luquillo – El Yunque complex) where silvicultural research, recreation, water catchment protection and wildlife protection, are successfully achieved in a national forest. Hunting and fishing in forest lands can be important in some areas of Central America although it is clearly different from the stereotyped image transmitted from the USA, Canada and European countries.

THE IMPACT OF VARIOUS TYPES OF MANAGEMENT ON THE PEOPLE LIVING IN THE FOREST OR ON ITS FRINGES: SOME CASE STUDIES

Management for protection

A series of studies have lately shown the considerable impact of soil losses affecting important watersheds (Paulet Iturri, 1978; Zerbe *et al.*, 1980; Wadsworth, 1978). Although precise data on soil losses are still scanty, and various estimates are given, generally speaking the appraisal is that this is a most severe problem. Erosion resulting from deforestation is considered the principal problem of Haiti, and Ewel (1977) analyzed the complexity of possible remedial action and the slim chances of achieving success. In the 1960s AID and USDA estimated that 80 per cent of the available land in Haiti was affected (Zerbe *et al.*, 1980). In Panama deforestation has caused sedimentation in Lake Alajuela—the main water storage to the canal system—to depths of 25 feet (7.6 m) in 1977 and by the year 2000, if present rates of sediment transport continue, about 40 per cent of the storage capacity will be lost (Wadsworth, 1978). With protection of the watershed the problem could be partly solved but the Panamanian government has been requested to solve the problem without relocating people (Zerbe *et al.*, 1980). In Honduras one USDA report (Zerbe *et al.*, 1980) stresses that

> in less than two generations at current rates of forest depletion, the forest cover will be virtually eliminated. The one-third of rural Hondurans who live on the edge of the forest will have no effective means for sustaining themselves through the life support system they presently use. In less than three generations, flooding in the valleys and alluvial plains, already serious, will be beyond the capacity of Honduras to manage with its own financial resources, and environmental damage will be irreversible.

The establishment and management of national parks has often resulted in effective protection by changing actual or potential destructive use to protective functions. Protection of water supplies has been one fundamental reason that helped to create a series of parks in Costa Rica and Panama, notably the

Volcan Poas National park (4000 ha) and the Chirripo National Park in Costa Rica (43,700 ha). More ambitious is the projected binational friendship park on adjacent territories in Panama and Costa Rica covering over 200,000 ha and including on the Panama side the projected Volcan Baru National Park, with another 30,000ha. Thousands of farmers who depend on the waters from these mountainous areas will benefit from these.

The attitudes of people living close to or in these areas, proposed for national parks, has been mixed. It is very difficult in some rural circles to understand that clearing for pasture or annual crops (usually a sequence beginning with timber exploitation, one or two years of annual cropping and ending in pasture establishment) cannot go on *ad libitum*. A series of contradictory moves by governments in the region can often be witnessed, responding to various types of pressures. Usually the official position is to condemn deforestation on steep slopes and/or on poor soils or with poor climatic condition (e.g. excessive rainfall). In practice, however, such policy is not sufficiently enforced and there is usually one branch of the government, usually linked with 'colonization' or 'agrarian reform' that is taking steps to provide new property titles to prospective farmers, often on lands that are inadequate for sustained cropping or grazing. Illegal grazing is another scourge that is a permanent source of conflict.

Only recently have efforts been undertaken to involve local people, either as workers in the national parks, by training some of the younger people as guides for visitors, or by providing fellowships for the children of inhabitants living close to the parks. Examples are reported from Honduras and Costa Rica. A workshop of park planners from various countries held in 1980 in the Cahuita National Park, Costa Rica (1100 ha in swamp forest, coconut beaches and 600 ha of coral reefs), stressed the need to consider park development as part of regional development. The group recommended that local people actively participate in the tourist bonanza and suggested a greater involvement of parks people in solving biological, social and economic problems of the people living in the area.

The main problem, however, remains that most people who have been traditionally farmers (including cattle ranchers practising extensive grazing), look at national parks, or protected watersheds, as a limitation to their traditional rights. The latter can be described as their right to use and abuse lands that have always been plentiful in the past. Squatters are still found in many national parks and it is extremely costly and sometimes dangerous to try to displace them.

The next few years will undoubtedly witness increased conflicts between governmental authorities and local people over issues raised by forest management for protection. It is a classical dilemma: how to weigh national interests against traditional individual customs or perceived rights. In some areas like El Salvador, Haiti or the Dominican Republic, overpopulation has so overstretched the carrying capacity that it is difficult to find workable alternatives. Each large storm (or hurricane) is likely to produce a new tragedy with loss of lives and property because lands have been abused in the past and the cumulative effect resulting in floods and erosion is likely to cause more severe calamities.

It is clear that no simple solution will work, and the subject lends itself to endless discussions among ecologically or economically oriented planners as well as social scientists. Possibly, some changes of land-use, involving various alternatives, should be considered, all of them aimed at relieving the pressure on land that should not be cultivated, grazed, or indiscriminately exploited. Suggested solutions involve the following alternatives, many of them already applied to varying degrees:

(1) intensification of land use for food production in promising areas and concentration of farmers on stable land-use systems in these areas;
(2) reforestation schemes coupled with an initial agricultural stage (for instance 'taungya') providing new jobs and a gradual change of land-use from destructive cropping or grazing to permanent forestry; and
(3) different schemes such as building of terraces and other measures to control erosion.

Some of these are not within the scope of the present chapter; others like agroforestry will be discussed below.

Forest management where wood harvesting is the main use

Natural forests.

In Central America there are still considerable areas under natural forests, which in countries like Panama, Costa Rica, Nicaragua and Honduras covers between 30 and 50 per cent of the land area. However, the forest cover is dwindling and this process has been described on a world and regional scale (UNESCO/UNEP/FAO/1978; US Department of State and USAD, 1978; Zerbe *et al*., 1980; Myers, 1980). For El Salvador and most of the Caribbean islands the amount of natural forests is very small and even what is left is often diminishing. Wherever natural forests are found, they are extremely important to local inhabitants who largely depend on them for food, poles, firewood, medicines, fibres and other products. Where shifting cultivation is practised the forest phase, provided it is sufficiently long, produces the regenerative effect that is needed to restore fertility or to break the destructive accumulation of weeds and insects that make permanent cropping impossible for low-capital, low-technology farmers.

In Central America most of the timber (and in the case of Honduras, pulpwood from natural pine stands) comes from these forests. Naturally this provides employment for local people but it often is not permanent because of the widespread practice of high-grading or 'mining' valuable species, leaving the remnant forest depleted, or more commonly, ready to be transformed into grazing lands. The latter stage itself may not be permanent either, especially in areas of high rainfall where fire cannot be used to avoid encroachment of woody vegetation in grasslands. The result is that grasslands that have replaced rainforests are often abandoned after only 5–20 years. The time depends on the degree of soil degradation, both chemical (nutrients) and physical (compaction) and/or the increasing invasion of weeds that cannot be controlled by hand-clearing because of rising costs. There is a large amount of secondary forests in areas that went through the phase of primary forest clearing – cultivation for 1–3 years – pasture establishment – abandonment – growth of secondary brush – secondary forest (sometimes with valuable species) – clearing again, etc. (Myers, 1980; UNESCO/UNEP/FAO, 1978).

The impact of these depleting processes is still little known. In some overpopulated countries like Haiti, El Salvador, the Dominican Republic, some lesser Antilles, Jamaica, the highlands of Guatemala and west Nicaragua, it undoubtedly causes hardship for local inhabitants, usually the poorest sections of the rural population. Particularly serious are the diminishing supplies of firewood from the forest, the various sources of food and construction wood and of medicinal products. It is likely that many of the internal migration processes can be explained by the disappearance of the forests. This is a subject that has been little studied.

An important aspect is the amount of hunting done in the forest by local inhabitants. In a speech at Yale University in April 1980, the present Director of Forestry of FAO, Dr Marco Flores, made the point that in his country, Honduras, over 80 per cent of meat consumed by rural inhabitants comes from hunting (mostly rodents, reptiles, and birds) on forest land. Other countries face similar situations and there is very little which is being done to ascertain the importance of hunting for food, let alone on how to manage wildlife for sustained yield.

A fundamental point that raises the question of permanent employment of people living within or close to the natural forest is that at present in Central America or the Caribbean, there is not one single example of a large area of natural primary forest that has been, over an extended period of time, managed as part of what may be called a well-conceived management plan. The same forest may have been exploited several times over the years, but not for the same species, and the end result has always been degradation. Perhaps this is a factor that must be given serious consideration when appraising socioeconomic impacts. Unfortunately, many decision-makers and foresters continue to insist that natural forest must be 'rationally' managed but they cannot show a single practical success case. Apparently, the more heterogeneous, the wetter, and the closer the forest is to primary conditions, the more difficult it is to manage tropical rainforests on a sustained yield for timber production. This implies that better prospects are found in associations where one or a few species dominate (certain swamp forests, tropical pine stands, etc.) and, of course, certain secondary forests. The latter present some excellent prospects for management in the wetter areas as has been shown in Costa Rica (CATIE, 1979).

One bright case can be found in the large Maria Tecum forest in Central Guatemala (over 2000 ha) close to the relatively large old city of Totonicapan (population over 50,000). In spite of the large demand for this highland forest (mostly various species of pines and one fir), the forest was able to resist over the last centuries the pressure for firewood, poles, timber and, in some less-timbered areas, the demand for sheep-grazing. This is essentially due to the fact that the land belongs to the municipality of Totonicapan, goverened by a council of elders with decisions that are very much obeyed by the predominantly Indian (Maya descendent) community. Cutting permits are given and are complied with, although they do not fulfil technical criteria from a forester's viewpoint. Obviously here is a case that deserves better study because it has what appears to be a good balance between local decision-making structures and compliance, a large number of people with decisions concerning valuable forest resources.

Plantations

Reforestation schemes have a considerable potential for benefiting rural people. At present they play a much smaller role than natural forests in the lives of the people but the ever-increasing scarcity of forest products is bound to change this. Moreover many reforestation schemes are also justified and funded because of the need to provide a protective cover for watersheds, particularly in wetter areas (at least this is the 'official' justification; it can of course be argued that protection against cattle poachers and fire can often bring about a perfectly good—and cheap—vegetation cover to prevent erosion).

At present there are large and apparently well-financed plans to reforest large areas of degraded lands in many countries of the region. Because of the scarcity of research and well-trained foresters many failures are likely to occur, concerning the choice of species, the selection of sites and, above all, adequate follow-up provisions to manage these plantations. However, considerable improvements will take place in the next few years and many possibilities of employment will result, hopefully not only connected with the preparation of sites and planting the trees, but also with the harvesting of the grown wood. Although there seems to be considerable interest in financing such reforestation schemes by lending agencies such as international banks and bilateral assistance programmes (Organization of American States,

1977), the impact on local populations is not well known because of lack of practical past experiences. At present, CATIE is carrying out a US $4 million research project funded by US/AID to ascertain the suitability of promising trees for firewood production for the six Central American countries. Initiated in 1980, the project covers 6 years and includes a sociologist. Many other schemes are being negotiated for Dominican Republic, Haiti, Honduras, Costa Rica and Panama.

It is interesting to notice that most of the reforestation is done on poor, often very degraded, sites. Lately however many trees have been planted, or at least tended, in areas with good soils; perhaps because of rising prices of timber and posts and economic incentives by governments. This has brought about a considerable upsurge of various agroforestry practices which will be considered below in more detail because of its promising possibilities and large impact on rural inhabitants.

Agroforestry

General considerations

As referred to earlier, agroforestry implies combining trees with crops or domestic animals, either simultaneously or in sequence. It is one of the curiosities of tropical forestry (and for that matter also in the agricultural sciences) that agroforestry has been rather blatantly ignored even if it existed all the time and played a most important role in the lives of rural inhabitants. With the exception of 'taungya', which is now well over 100 years old, very little is known concerning most agroforestry systems, even if some of them appear to be extremely productive and rather sophisticated—and of course stable—possibly as a result of evolution over decades or centuries. At present much of the research in various countries amounts to the 'discovery' and description of existing systems. It is interesting to speculate why this is so, and many arguments have been advanced: the avoidance by scientists of agroforestry systems because of their complexity and the great difficulty in setting up an experimental design; the present training of scientists in specialities rather than in complex land-use systems (how much do tropical foresters know about food plants and how do agronomists fare vice-versa?); the past reliance on cheap direct or indirect energy sources in agriculture, able to replace the beneficial effects of trees on crops (fertilizing, weeding, ploughing, pest control, etc.); and perhaps more than anything else, the indiscriminate transference to the tropics of capital and technology-intensive methods that originated in temperate regions. Finally the socio-cultural dimension of agroforestry, so important in many countries, adds still another dimension that complicates a better understanding of the problems, at least from the researcher's viewpoint.

Moreover, there are a few mental blocks that must be overcome concerning agroforestry. Among them there is the belief that agroforestry is mostly practiced by the rural poor or that it is restricted to 'marginal' land (see for instance Table 1). This has been particularly illustrated in the publication that led to the foundation of the International Council for Research in Agroforestry, ICRAF (Bene *et al.*, 1979) which stated that '. . . more than half of all land in the tropics, although too dry, too steep, too rocky to be classified as arable land, is suitable to the practice of agroforestry', and this statement in fact, was placed in an emphasis box (p. 43).

Actually the more agroforestry is studied, the more it becomes clear that it has achieved productive and stable systems on all kinds of lands, on good and poor soils, with varying relief, climatic conditions and under different socio-economic conditions. Students that have taken agroforestry courses

Table 1 Comparison of advantages and disadvantages of some agroforestry systems (presence of trees within croplands or pastures) as compared with monocultures of these same crops or pastures

Advantages	Disadvantages
BIOLOGICAL AND PHYSICAL ASPECTS	
A better utilization of the vertical space and cropping period is achieved and natural ecological models are simulated as to form and structure: solar energy is more efficiently captured.	Trees compete for light needed for crops or grass.
A larger biomass returns to the system (organic matter) and often is of better quality.	Trees may compete for nutrients.
There is more efficient recycling of nutrients including their 'pumping' from the deeper soil layers.	Trees may compete for water (especially where there is a strong dry season).
Appropriate for marginal areas since there is likely to be a larger resistance to rainfall variability; it can also be practised on steeper slopes.	There may be allelopathic influences.
The damaging action of rainfall and wind is reduced.	The harvesting of trees may cause damage to crops.
Fertilizers applied may show effects for a longer time since tree roots can capture nutrients that move beyond the area of crop or grass roots and recycle them.	There is no 'rest period' (with a secondary forest stage) like in shifting cultivations.
Leguminous trees (and some from other families) fix and incorporate nitrogen.	Mechanization is impossible or made more difficult.
There is more mulch and lesser weed growth.	The greater air moisture can favour diseases (mostly fungi).
	Raindrops forming in the crown of trees can cause damage.
	A proliferation of noxious animals can be favoured.
	There could be excessive export of nutrients if trees are consistently harvested for sale outside the area.

Table 1 (*continued*)

Advantages	Disadvantages
The soil structure is improved (more stable aggregates) and a hard pan is prevented.	
Diversity of species hampers insect proliferation.	
ECONOMIC AND SOCIAL ASPECTS	
The farmer (particularly poorer ones) satisfies, partly at least, his needs for firewood, poles, posts, timber, food, flowers for honey, medicinal products, etc., and does not need to buy these.	The system is more complex and little understood. Any experimental design with statistical analysis is likely to be more complicated.
Trees constitute a 'standing capital', an assurance against emergencies, when quick cash is needed.	In some cases yields of crops or grasses are smaller with lower economic returns.
Dependency on one single crop is avoided and traumas associated with irregular rainfall, market fluctuations, dependency on imported products (pesticides, fertilizers), outbreak of pests, are mitigated.	In many ongoing economic structures it is considered that the system is not 'efficient' because of the inherent complexity of operations needed to make it function effectively.
There is less need to 'import' energy and/or to pay for products outside the area.	Normally, more labour is required.
Investments to establish tree crops are reduced (such as in 'taungya').	It is argued that some agroforestry practices do not stimulate farmers in moving out of their present socio-economic status (poverty or subsistence level).
Diversity and the interdisciplinary nature can be considered an asset for quality of life.	Economic recuperation may take a longer time for people with low incomes (because of the time lag in cropping trees).
Social cohesion and teamwork can be promoted.	There may be opposition from some political and socio-economic structures towards agroforestry systems because of ignorance or false premises.
Trees serve to mark property boundaries and constitute a safeguard device against land usurpation.	There is scarcity of trained personnel to improve and handle agroforestry systems and install new ones.
Some schemes allow a gradual change from destructive land-use practices towards more stable systems, without diminishing productivity.	There is lack of knowledge on the potentialities of agroforestry systems among decision-makers and this results in scarcity and lack of funds for research and extension programmes.
There is more opportunity to distribute labour during the year.	
Wildlife is favoured and in some instances can be a suitable source of protein.	
There is obviously considerable scope to design new systems by associating the most desirable species.	

at CATIE report that they are now 'discovering' agroforestry systems in areas they had previously visited repeatedly without noticing them.

A full discussion on agroforestry practices in Central America and elsewhere in Latin America can be found elsewhere (Budowski, 1977, 1979 and 1980; CATIE, 1979, Combe and Budowski, 1979). The present discussion will only attempt to emphasize some examples with a few speculations on social implications. A good introduction to the subject would be a comparison between agroforestry systems and monocultures resulting from lengthy discussions and exchange of experiences between researchers in Mexico and CATIE. Table 1 should be considered as tentative and many of the statements found are clearly subjective. They represent mainly opinions or viewpoints by different people, not necessarily based on proven measurements and valid comparisons, aimed above all to stimulate discussions. It should be noticed that the second part of the table emphasizes socio-economic implications.

A few successful case studies of agroforestry

Alder in the highlands of Costa Rica. For at least 80 years a local alder species (*Alnus acuminata*) has been successfully planted at elevations between 1300 and 2500 m with high rainfall (2000–3000 mm) and good soils in the dairy region in Costa Rica. The presence of alder trees is said by many to increase fodder production. This is an area where milk production is high and many farmers are relatively affluent. The trees are planted at wide spacing within pastures that are grazed (*Pennisetum clandestinum*) or used for cutgrass (*Pennisetum purpureum* and *Axonopus scoparius*). Most alder trees found in pastures are regularly pruned. The spacing of the alders can be any combination between 8 × 8 m and 15 × 15 m. Alder fixes nitrogen through large nodules (*Frankia* sp. of the actinomycetes). Some initial measurements indicate a mean annual diameter increment of 2–3 cm with harvesting after 15–20 years (CATIE, 1979). The easily worked wood is used for multiple purposes where durability is not a requisite. The social implications have been partly studied by Poschen (1980) through questionnaires distributed to dairy farmers, and reveals a series of beliefs and assumptions among farmers that deserve more careful research. The practice of planting alder in pastures is at present increasing.

Coffee or cocoa with one or two strata of 'useful' shade trees. Although the trees above coffee or cocoa are usually referred to as 'shade trees', their function and usefulness is clearly much greater than strictly for shade. Many of these advantages are outlined in Table 1. The system over coffee and cocoa has evolved over the past years and it is now common in various regions, particularly in coffee, to have a leguminous tree as an intermediate canopy

(*Erthrina* spp. or *Inga* spp.) overtopped by a tall and valuable timber tree with a narrow crown (e.g. *Cordia alliodora*). The leguminous nitrogen-fixing trees are severely pruned back, usually twice a year. Several plots of *Cordia alliodora*, 15–17 years old, showed a mean annual diameter increment of 1.8 cm, an average height of 22–23 m and a mean annnual volume increment of 10–11 m^3/ha (Combe and Gewald, 1979). The practice is increasing and several measurements show that coffee yields are high and compare favourably with neighbouring plots without *Cordia alliodora*.

Life fenceposts that produce wood, food, feed and more fenceposts. Throughout tropical America the practice of raising trees from large cuttings that root easily, to establish fences to keep cattle out, is extremely widespread. Fences can be grown by planting cuttings closely together, often mixed with low brushes or, more commonly, they serve to attach barbed wires. These practices have received very little literature coverage and they are currently being investigated as to their biological, social and economic effects. The summarized analysis (Table 2) again should only be considered as preliminary since it covers over 50 tree species in six Central American countries over a great diversity of conditions.

Table 2 Comparison between live fenceposts and wooden fenceposts (the latter either naturally durable or treated with preservatives); both posts are being used to attach barbed wires and to prevent trespass, principally by cattle

Factor	Live fence	Wooden (dead) fence
Choice of species	Depends on ecological conditions	Many possibilities: depends on availability
Cost	Relatively low or free	Relatively high
Handling of post before placing	Needs careful preparation, transport and storage	No special care required
Placing in soil	Needs care, adequate soils	Soil not limiting
Placing of barbed wire	Special techniques in some species	Some skills required
Initial maintenance	Necessary, requires protection against some animals	None, in some cases needs fire protection
Survival	Losses possible	100 per cent
When to place wire	Usually when well-anchored	Immediately
Increase of post density along fence	Easy and cheap	Easy by expensive

Table 2 (*continued*)

Factor	Live fence	Wodden (dead) fence
Durability	Usually very long	Variable, limited according to treatment and species
Organic matter production	Varies with species	None
Nitrogen-fixation	Possible in some species	None
Effect on soil fertility	Beneficial, specially when branches are pruned and some roots die off (aeration)	None
Erosion control	Can be effectively used as barrier	None
Competition for water and nutrients and light with nearby crop	Does exist but varies according to system; organic matter production compensates	None
Protection of crops and/or animals against wind	Effective but varies according to species height, density	None
'Horizontal rain' (fog drip) from moisture-laden winds	Possible	None
Toxic effects	Possible (allelopathy)	None (except when some preservatives are used)
Harmful fauna	Can be sheltered	None (except termites in some cases)
Beneficial fauna	Provides shelter and food (e.g. birds, honeybees)	Little
Additional economic products	Many, such as food, feed, medicinal products; also firewood, posts and more live fenceposts	None
In case removal is necessary	Difficult and costly	Relatively easy
Labour for management	Periodical pruning is necessary; skills required	Skills also required to place posts and wires and replace them
Acceptance by farmer	Very popular among poorer farmers	Depends on income; more affluent farmers tend to avoid live fenceposts
Special limitations	Disliked by fumigation pilots	Firebreaks must be kept clean during fire season
Aesthetic appreciation	Depends on management and cultural background	Depends on investment possibilities and cultural background

Taungya as part of a strategy to change land-use. Throughout Central America and the Caribbean a large amount of land on steep slopes has been cleared of its original forest and converted to poor pastures, many of the latter reverting eventually to secondary brush. At present there appears to be great interest in converting such slopes into forest plantations for firewood, pulpwood or timber or any possible combination. Taungya has a long history in the region, mostly in Trinidad (as early as the 1920s), with teak and Caribbean pine; Belice (teak, Caribbean pine, *Gmelina*); Surinam and at CATIE, Costa Rica, the latter with *Cordia alliodora, Gmelina arborea, Eucalyptus deglupta* and *Terminalia ivorensis* (Combe and Gewald, 1979). The latter four species and *Pinus caribaea* have shown particularly good adaptation in areas of high rainfall (2000–4000 mm) at elevations up to 1200 m. The associated crops have been maize, beans, cowpeas and string beans.

A series of programmes are at present being considered, the aims of which are to convert large areas of degraded lands in all of the Central American countries, as well as Dominican Republic and Haiti, into tree plantations. Thousands of farmers are expected to be involved in the process by allowing them to combine field crops with seedlings of valuable trees for approximately 2 years and then having them participate in the tending and harvesting operations, therefore changing gradually land-use from extensive (and destructive) grazing to forestry. The rising prices of firewood, posts, timber and pulp, and the possibilities of eventually producing large and cheap sources of biomass on lands that do not compete with agricultural crops, have generated much interest from funding agencies. In Costa Rica the introduction of Caribbean pine 20 years ago, and the ready acceptance of local farmers to plant small plots, has led to the establishment of pines and other species. The eventual grazing of these plantations under controlled conditions is being studied, with a view to keeping the grass down, reducing weeding costs and giving additional income. There is already some experience on grazing pine plantations in Surinam and Jari (Brazil) and field trials are under way at CATIE, Costa Rica.

The obvious interest in financing such schemes arises from a combination of factors: protecting degraded slopes from erosion and regulating run-off, providing wood for an industry that is willing to pay higher prices, to provide firewood and charcoal which are in great demand, but above all because it would provide thousands of farmers—who have become impoverished from past misuse of land—with a new source of income, without drastically changing their way of life. Clearly all these prospects need to be investigated in the light of land-tenure problems, government incentives, long-term economic prospects, acceptability by the farmers, and careful planning of the relationship between wood-using industries and participating local farmers. There is an obvious role for governmental authorities to play but misunderstandings and failures can easily happen.

Multiple use

The possibilities of promoting multiple use of forest land, particularly as a medium to improve the lives of people living near or within the forest, have not yet been well investigated in this region although this may not be so much a result of insufficient case studies but rather not considering adequately the social factors.

A case in point is the Luquillo National Forest in East Puerto Rico covering an area of low and middle elevation mountain ranges, with heavy rainfall. About 40 years ago the forested area, mostly secondary, was dedicated to silvicultural research (managing the natural forest and establishing plantations) as well as watershed protection, but increasingly over the years it has become a main focus of attraction for local and foreign tourists with recently over 1 million visitors per year. The effect on the local nearby population has not been well studied but it certainly has produced a powerful attraction to the Puerto Rican tourist industry with various favourable economic implications.

Another interesting development has been the creation in 1979 of the Rio Platano biosphere reserve in Northeastern Honduras, covering 180,000 ha of rainforest (Digerenare and CATIE, 1978). Besides the biological and archaeological interest for science, a prominent reason for the establishment of the reserve has been the maintenance of the lifestyle of the well-adapted 'Misquito' with their peculiar agricultural, hunting and fishing practices.

A better knowledge of the relation between multiple use and its influence on people living in the area where practised is needed. The subject is wide open for research, and the results would be of great use to planners and decision-makers.

SUMMARY

Four broad categories of forest management in Central America and some countries of the Caribbean—namely for protection, for wood production, for a combination of agriculture and forestry (agroforestry) and for multiple use—are analysed as to their present and potential impact on local populations living within or near the forest. Many problems resulting from past and present massive deforestations and lack of management are adversely effecting local populations. In the light of experience from the region it appears that because of the shortages of forest products and the need to control erosion, large reforestation schemes on land that has been degraded would provide interesting opportunities for stable forest communities. Capitalizing on present widespread agroforestry practices appears to be particularly promising in some areas but a better understanding of biological and social characteristics of some of the present systems is required to launch successful improvement programmes. Some examples of multiple-use forestry are also described.

REFERENCES

Bene, J. G., Beall, H. W. and Cote, A. (1977). *Trees, Food and People, Land Management in Tropics.* Ottawa, IDRC.

Budowski, G. (1977). 'Agro-forestry in the humid tropics. A programme of work'. Report submitted to IDRC, Turrialba, Costa Rica, CATIE.

Budowski, G. (1979). National, bilateral and multilateral agro-forestry projects in Central and South America. In *International Cooperation in Agro-forestry*, 15–23 July 1979, Nairobi, Kenya. (in press).

Budowski, G. (1980). The place of agro-forestry in managing tropical forest. In *International Symposium on Tropical Forests: Utilization and Conservation* (ed. F. Mergen). Yale University, New Haven, USA, 15–16 April . (in press).

CATIE, (1979). *Workshop on Agro-forestry Systems in Latin America,* Turrialba, Costa Rica 26–30 March. Proceedings.

Combe, J. (1979). Combinaciones silvo-pastoriles en zonas altas de producción de leche en Costa Rica. In *Proceedings Simposio Internacional sobre las Ciencias Forestales y su Contribución al Desarrollo de la América Tropical*, 11–17 October 1979. Notas para la excursión No. 3, (in press).

Combe, J., and Budowski, G. (1979). Classification of agro-forestry techniques. CATIE (1979) pp. 17–47.

Combe, J., and Gewald, N. (1979). *Guía de campo de los ensayos forestales del CATIE en Turrialba.* CATIE, Turrialba, Costa Rica.

Digerenare and CATIE (1978). *La cuenca del Río Plátano (Mosquitia Honduras).* CATIE, Turrialba, Costa Rica.

Ewel, J. (1977). 'A report on soil erosion aid prospects for land restoration in Haiti'. Submitted to USAID, Port-au Prince, Haiti. Contract AID/la-C-1196. (mimeo).

ICRAF (1979). Newsletter; news and comments from the *International Council for Research in Agro-forestry,* **1** (1), 1–4.

Myers, N. (1980). *Conversion of Tropical Moist Forests.* National Academy of Sciences, Washington DC.

Organizacion of American States (1977). 'Plan de acción para el desarrollo regional de la línea noroeste (República Dominicana)'. OEA, Washington, DC.

Paulet Iturri, M. (1978). *La erosión y la conservación de los suelos en la República Dominicana.* IICA, Sto. Domingo.

Poschen, P. (1980). *El jaúl con pasto. Anzatze eines silvo-patorales Systems auf Viehweiden der submontanen Stufe in Costa Rica.* Diplomarbeit, University of Freiburg (Fed. Rep. of Germany), Faculty of Forest Sciences.

UNESCO/UNEP/FAO (1978). *Tropical Forest Ecosystems; A State of the Knowledge Report.* UNESCO, Paris.

US Department of State and USAID (1978). *Proceedings of the U.S. Strategy Conference Tropical Deforestation*, 12–14 June. Washington, DC.

Wadsworth, F. (1978). 'Deforestation – death to the Panama Canal'. In *Proceedings of the U.S. Strategy Conference Tropical Deforestation*, 12–14 June, pp. 22–4. Washington, DC.

Zerbe, J. II, *et al*. (1980). *Forestry Activities and Deforestation Problems in Developing Countries.* Report to AID, PASA No. AG/TAB-1080-10-78, Washington, DC.

Socio-economic Effects and Constraints in Tropical Forest Management
Edited by E. G. Hallsworth

10

The agroforestry system at Mae Muang Luang, Thailand

PISIT VORAURI
Faculty of Agriculture, Chiang Mai University, Chiang Mai, Thailand

INTRODUCTION

Serious environmental and socio-economic problems in Thailand are related to the decreasing forest area. In 1963, the total forest area was 51 per cent or 255,000 km^2, and in 1973 it had decreased to only 38 per cent. However, since 1973 the alarming rate of forest destruction higher than 3 per cent per annum was recorded and in 1977 only 25.35 per cent of undisturbed forest was left in the country. Reduction of the forest area is closely associated with farming in the highlands. In the north-east, east and west, claimed lands are used for cultivation of maize, tapioca and sugarcane. After 2–3 years of cultivation without proper soil management the land becomes infertile. In the north the situation is even more complicated; cut fields on steep slopes in undisturbed forest are used for subsistence cropping in the rainy season and for illicit growing of the opium poppy in winter. Generally, sheet erosion in the first and second cropping seasons is clearly noticeable. Usually on a slope between 30–45 per cent 20–30 cm of topsoil is eroded during these two seasons. This results in the land being abandoned and later Imperata grass is established and becomes the grazing ground for the highlanders' cattle. In the dry season some of the highlanders initially induce forest fires to the dry and highly inflammable Imperata grassland. From this small starting point a forest can spread miles. This has been done year after year in order to force the Imperata grass to produce new growth for the cattle.

The renewal of the forest by nature is only a dream. The foresters realize the impact of the shifting cultivation on environmental and socio-economic problems, but very few of the rest of the country population realize this.

In spite of the Reforestation Programme which has been conducted by the Royal Forest Department for decades, the programme itself has up to now consisted only of planting forest trees in Imperata grassland areas. However, the highlanders still shift from place to place continually and cut fields in the undisturbed forest so the practice of shifting cultivation is still employed. Fortunately, in the last 5 years the Royal Forest Department have set up a Forest Village Programme in which the highlanders are encouraged to established permanent villages and work on permanent farming in the designated areas. Another agency besides the Royal Forest Department, the Forestry Industry Organization, is also responsible for reforestation in certain areas. This organization has employed the modified taungya system of Agroforestry in their reforestation programme since 1968. Shifting cultivators were collected and put to settle in villages; they were offered work in the reforestation programme and allowed to grow agricultural crops between the rows of the forest trees while trees were still low. This programme achieved a certain success and it was suggested that some other kinds of approach should be initiated in order to stop shifting cultivation.

AGROFORESTRY SYSTEM AT MAE MUANG LUANG

The activities concerning agroforestry research and extension work at Chiang Mai, Thailand were initiated in January 1979 in the Huai Thung Choa Watershed Improvement Unit as one of the United Nations University-sponsored activities under the Natural Resources Programme. The actual research site is at Mae Muang Luang, 17 km west of Huai Thung Choa and 90 km north-west of the city of Chiang Mai. The main objectives of the agroforestry programme are:

(1) to study the possible socio-economic and ecological impact of agroforestry in northern Thailand;
(2) to introduce agroforestry systems to concerned people, i.e. administrators and academic staff as well as farmers, both in the highlands and the lowlands.

The programme was planned on the basis of four assumptions. First, the highlanders do not farm on land which has more than 30 per cent forest cover, as they normally clear and burn before planting. Second, communication by road between remote villages is a major problem for the highlanders. Third, the supply of food, especially rice, becomes very short during the rainy season; thus any agroforestry system must include rice cultivation. Fourth, if trees are to be grown in potential swidden areas, the reason should not be simply that they are grown for firewood or for construction materials. Other benefits should include soil protection, shade, and soil regeneration.

Experiments

The first experimental plots were designed and set up in mid-1979. The 27 experimental plots of 0.32 ha each have been established on slopes of 10–15 degrees. The design of the plots is based on a three by three matrix with three replicates. Three groups of crops are grown in varying proportions; namely forest trees, perennial crops (coffee) and annual crops (primarily rice). In 1980 the researchers planted a variety of annual crops in narrow strips along the contour line. The annual crops tested included upland rice, green cabbage, Chinese radish, digitalis, lemon grass, Mung bean and greenleaf desmodium. Superimposed on these test plots is an experimental design to elucidate the effect of different levels of fertilizer application on the coffee trees within the agroforestry plots. Extensive testing of the chemical and physical properties of the soil in the test plot area has been carried out to provide a basis for long-term comparisons.

A second group of experimental plots was established in 1980, and this consists of five combinations of perennial crops as follows:

(1) coffee, forest trees (*Erythrina* sp.) and temperate fruits,
(2) coffee and forest trees;
(3) fruit trees and two species of forest tree (*Betula alnoides* and *Erythrina* sp.)
(4) mango, *Erythrina, Eucalyptus, Pinus caribaea* and *Cordia alliodora*;
(5) coffee, *Gmelina arborea, Melia* spp.

Results

At this stage of the experiment it is too early to assess clear-cut results from the study. However the results have already shown some trends of these aspects as follows.

Agronomy aspect

Some results indicated suitable growing requirements of the plants experimented on in different shaded areas from 0, 15 up to 30 per cent using different kinds of forest trees. It was interestingly found that rice, Chinese radish and Chinese mustard gave a lower yield when grown under a more shaded area. It could also be observed that the rice yield was also related to the slope of the plot. When the slope was increased the yield conversely decreased.

Socio-economic aspect

It is still not possible to evaluate the results of the socio-economic aspect at this stage. However, from general observation it appeared that those highlanders who had better social and economic status more easily accepted the system when it was introduced to them, whereas those who were barely making the miniumu rate for subsistence were still waiting to see the results, and continued to use the traditional system.

It showed evidently that the highlanders did not accept the concept of growing forest trees either for firewood or for construction purposes. This is because wood for both purposes can still be found in natural forests next to their villages. Moreover, growing annual crops in a plot having more than 5 per cent of forest trees in the area is still unacceptable. Nevertheless, the concept of growing trees to enrich the soil as well as for shading purposes, for example *Erythrina*, has begun to be acceptable among the highlanders.

IMPACT

Setting up the experimental plot on agroforestry at Mae Muang Luang has had a direct effect on the farming of the highlanders; for example, they started to grow coffee under the shade of forest trees growing naturally; they also planted rice, maize as well as green legumes along the counter lines and made drainage ditches on the contour line every 20–30 m instead of growing these crops conventionally on the slopes. Moreover, they also planted each kind of crop in strips in place of mixed cropping as before.

It should be mentioned that the foresters who are in charge of watershed conservation are of the opinion that one of the methods to prevent forest destruction by swidden farming is to work on the management of watersheds and their inhabitants by introducing a permanent agroforestry system into the area.

Socio-economic Effects and Constraints in Tropical Forest Management
Edited by E. G. Hallsworth

11

Socio-economic thrusts in an integrated forest management system: the Philippine case

CELSO P. DIAZ
Socio-Economics Research Division, Forest Research Institute (FORI), Ministry of Natural Resources, College, Laguna 3720, Philippines

INTRODUCTION

Located in the wet tropical zone between 5° and 21°N latitude and 117° and 127°E longitude, the Philippines is an archipelago of some 7100 islands on the western rim of the Pacific and stretches more than 1000 km north to south between Taiwan and Borneo. It is bounded by the Pacific Ocean on the east, the Celebes Sea on the south, the Sulu Sea on the south-west and the Balintang Channel on the north. The country has a total land area of 30 million ha and a total coastline of 17,461 km (PCSS, 1980). The three biggest islands are Luzon, Visayas and Mindanao. In 1979 the population was estimated at 47 million and accordingly if an annual growth rate averaging 2.7 per cent has to continue, as projected by the National Census and Statistics Office (NCSO), the population by the year 2000 will reach a total of nearly 84 million people. Around 67 per cent of the total population live in the rural areas (71 per cent in 1975). Projections using urban, rural and natural growth rates indicate that by the year 2000 the urban population will match that of the rural areas.

The Philippines has a generally mild tropical climate. It is characterized by a relatively high temperature, high humidity and abundant rainfall. The rainfall throughout the year ranges from 150 to 450 cm while the mean annual temperature is 27°C. During the months from July to December typhoons frequently occur, often aggravated by the monsoon rains which cause the flooding of low-lying areas that damages foodcrops, other agricultural products, infrastructure and even human lives.

The Philippine Constitution declares that all the natural resources of the country belong to the state. This embraces all lands of the public domain, waters, minerals, coal, petroleum and other mineral oils, all forces of energy, fisheries, wildlife and other natural resources. It further classifies lands of the public domain into agricultural, industrial or commercial, residential, resettlement, mineral, timber or forest, and grazing lands, and such other classes as may be provided by law. For the exploitation and development of forest resources the Constitution also provides that no private corporations or associations may hold by lease, concession, licence, or permit, timber or forest lands and other timber or forest resources in excess of 100,000 ha; although such area may be increased by the National Assembly upon recommendation of the National Economic and Development Authority.

THE FOREST RESOURCES

In terms of forest resources, the Philippines grows some 3800 wood species, 300 of which are considered indigenous (Pollisco and Natividad, 1980). Non-wood resources such as rattan, bamboo, palms, vines, orchids and others also abound within the forest lands. It is commonly believed that for a major portion of the Philippine archipelago, the climax vegetation was forest. Subsequent significant changes in land cover may be attributed largely to the work of man through cultivation, grazing, logging and other land-uses.

Forest land-use

At the end of the 1970s (December 1979), forest lands covered about 55 per cent or 16.91 million ha of the total land area of the country (Table 1). Of this total, 9.33 million ha or 55 per cent had been classified into the following broad land-use categories:

(a) forest reserves and timberland	8,630,000 ha
(b) national parks	316,117 ha
(c) military reservations	129,911 ha
(d) civil reservations	252,524 ha

The balance of 7.58 million ha or 45 per cent are still unclassified forest lands of which 4.3 million ha have been programmed and predetermined to be classified as alienable and disposable. These areas are currently the subject of nation-wide classification by government composite land classification teams. Of the total area of 16.91 million ha of forest lands, 11.2 million ha or 66 per cent are covered with timber, while 5.7 million ha or 34 per cent are already devoid of timber (Table 1). Luzon and Mindanao account for the bulk of forest lands in the county with 39.67 per cent and 37.21 per cent respec-

Table 1 Land-use status of the Philippines as of 1979 ('000 ha)

Forest type and stand size	Forest land (16,907)	Alienable and disposable (13,093)	Total (30,000)
I. *Forest*	*11,175*	*1,486*	*12,661*
(A) Productive	9,445	1,456	10,901
(1) Dipterocarp	9,034	1,427	10,461
(a) Rep-brush	2,671	937	3,608
(b) Young growth	3,274	388	3,662
(c) Old growth	3,089	102	3,191
(2) Mangrove	218	27	245
(a) Rep-brush	104	19	123
(b) Young growth	103	7	110
(c) Old growth	11	1	12
(3) Pine	193	2	195
(B) Unproductive/protection	1,730	30	1,760
(1) Dipterocarp	1,398	24	1,422
(2) Mossy	330	1	331
(3) Bamboo	2	5	7
II. *Non-forest*	*5,732*	*11,607*	*17,339*
(A) Open land	426	450	876
(B) Managed pasture	967	27	994
(C) Marsh and small water	77	41	118
(D) Plantation upland	1,921	5,125	7,046
(E) Cultivated cropland	2,116	5,361	7,477
(F) Urban and others	225	603	828

Source: Philippine Forestry Statistics. BFD, 1979.

tively, while the remaining 16.31 per cent and 6.81 per cent are in Visayas and Palawan, respectively. Likewise, Luzon and Mindanao get the biggest share of the total alienable and disposable lands of 13.1 million ha with 41.25 per cent and 29.85 per cent respectively. The administration and management of the entire forest lands is within the function of the Bureau of Forest Development (BFD).

The objective of the present system of land classification is to retain 42 per cent of the total land area (30 million ha) of the country for forest land purposes and the remaining 58 per cent available for agriculture, industrial, residential and other uses (BFD, 1979). In order to attain this objective, all lands 18 per cent (10°) in slope and over are to be retained for forest purposes. Only areas below 18 per cent in slope can be classified as alienable and disposable, provided they meet other criteria set by law. However, where the public interest so requires, areas previously classified as alienable or disposable may be reverted to the category of forest land.

Size, type and location of the forests

The Philippine forests are one of the most diversified in tropical south-east Asia. A great variety of tropical broadleaved hardwoods, of which the Dipterocarp group predominates, as well as two species of pines, are represented. Of the over 2000 species of trees that attain a diameter of 30 cm or more, only the woodworking qualities of less than 200 species are known.

In terms of commercial forest, the country has 7.2 million ha (including young growth and old growth), of which 6.7 million ha or 93 per cent are within forest lands while 451,000 ha or 7 per cent are in the alienable and disposable lands. Geographically, of the 6.7 million ha mentioned above, 3.6 million ha or 53.37 per cent are in Mindanao, 1.9 million ha or 28.34 per cent in Luzon, 66,000 ha or 9.89 per cent in the Visayas and 56,000 ha or 8.40 per cent in Palawan.

In general, forest vegetation in the Philippines is classified into five main types as follows:

(1) Dipterocarp type—a mixed tropical rainforest with Dipterocarps as the predominant commercial tree species. This type is the most important from the standpoint of production and trade. This occupies about 9.03 million ha or 65.6 per cent of the total productive forest lands of the country. Practically all major industries dependent on wood as raw materials—from logging to wood processing—are based on Dipterocarp forests. Harvestable timber from this stand averages between 90 and 150 m^3/ha. Among the trees in this type are the world-renowned Philippine mahogany species.

(2) Molave type—this forest is characteristically less dense and more open than any other types, and may be found in areas of poor calcareous soils and of relatively dry climate. Occurring either in isolated patches or blending with the other forest types, woods of the few species, notably *Dracontomelon dao*, *Pterocarpus* spp. Linn., and *Intsia bijuga* are highly valued for their natural beauty and durability.

(3) Pine-type—the pine forest occupies the highland regions of mountain ranges in northern Luzon. It is mainly composed of *Pinus kesiya* and *P. merkusii*. The former is extensively used for mining purposes in northern Luzon while the latter species, which occupies a smaller area, is seldom used. The species are also tapped for their resins, and utilized as timber and watershed cover where they occur naturally. This type covers an area of 193,000 ha or about 2 per cent of the total productive forest lands.

(4) Mangrove type—this type occurs on tidal flats bordering coastal areas, on the shores of protected bays and along the mouths of streams where water is brackish. This forest is the main source of tan bark for the tannin extract industry and the wood is used for fuel and charcoal-making. In previous years, sizeable quantities of wood were exported for rayon.

Main species are *Rhizophora* spp. and *Brugiera* spp. It has an area of 218,000 ha or about 2.3 per cent of the total productive forest land.

(5) Mossy type—this type consists of dwarf-like tree vegetation clothed with clinging mosses and liverworts and usually found in high and very mountainous regions. This forest is valued more for protecting watersheds than for its commercial value. It has an area of 330,000 ha or about 19 per cent of the total protection forest within forest lands.

Timber resources

The productive forest, as shown in Table 1, has a total area of 9.44 million ha which includes old-growth forests, adequately stocked residual or young growth and areas of low density stands. The old growth forest with an estimated volume of 830.44 million m^3 (Table 2) will be the major source of wood until such time as the young-growth forest will be ready for harvest. Prediction by a local forest management expert (Revilla and Bonita, 1978, as cited by Pollisco and Natividad, 1980) revealed that the present old growth may last only until the year 2000 if a cut at the rate of 100,000 ha annually is maintained with an average production of 10 million m^3 of timber per year. The government hopes by then that the young-growth forest will be ready for the second cut.

Watershed areas

Within the forest lands there are 40 proclaimed watershed reservations, consisting of about 351,000 ha. Some of them are presently supporting big hydroelectric power dams while the rest are all potential sites for geothermal explorations and irrigation projects. These watersheds, however, are in danger due to improper land use such as kaingin-making, illegal logging, burning and other forms of destruction. Due to these uncontrollable practices the areas have been subjected to excessive soil erosion (Table 3). This situation threatens the life of many government financed multi-million dollar infrastructures in the country. Hence, the government needs to adopt urgent measures to abate the further destruction of those watersheds. Present estimates show that at least 1 million ha are considered critical within watershed areas that need immediate rehabilitation.

National parks/wildlife sanctuaries

There are 60 national parks and seven wildlife sanctuaries which have been established with an aggregate area of 2,015,067 ha. By law these areas are closed to any form of exploitation but through the years encroachment seems uncontrollable despite the many laws enacted to protect them.

Table 2 Volume of standing timber in forest lands by type, stand size and geographical region, 1979 ('000 m^3)

Forest type and stand size	Philippines		Luzon		Visayas		Mindanao		Palawan	
	Volume	Percentage distribution	Volume	Percentage distribution	Volume	Percentage distribution	Volume	Percentage distribution	Volume	Percentage distribution
TOTAL	*1,586,728*	*100*	*408,163*	*26*	*179,594*	*11*	*867,519*	*55*	*131,452*	*8*
I. *Productive forest*	1,464,863	100	378,871	26	130,989	9	846,477	58	108,526	7
(1) Dipterocarp	1,440,565	100	361,426	25	130,451	9	843,006	59	105,682	7
(a) Rep-brush	75,727	100	36,815	49	16,500	22	18,465	24	3,947	5
(b) Young growth	534,494	100	112,539	21	42,424	8	363,446	68	16,085	3
(c) Old growth	830,344	100	212,072	26	71,527	9	461,095	55	85,650	10
(2) Mangrove	7,393	100	540	7	538	7	3,471	47	2,844	39
(a) Rep-brush	1,178	100	50	4	510	43	210	18	408	35
(b) Young growth	4,745	100	490	10	28	1	2,829	60	1,398	29
(c) Old growth	1,470	100	—	—	—	—	432	29	1,038	71
(3) Pine	16,905	100	16,905	100	—	—	—	—	—	—
II. *Unproductive forest*	121,865	100	29,292	24	48,605	40	21,042	17	22,926	19
(1) Dipterocarp	106,771	100	27,263	25	46,883	44	15,946	15	16,729	16
(2) Mossy	15,094	100	2,029	13	1,772	12	5,096	34	6,197	41

Source: Bureau of Forest Development, 1979.
Note: The data above were obtained in the nationwide forest inventory conducted in 1962–68 by the BFD with USAID, and PAF assistance.

Table 3
(A) Sediment load in selected river basins in the Philippines (1978)

River	Watershed area (km²)	Average anuual sediment load (tons/year)	Average annual sediment load (tons/km² annually)
1. Talavera	261	577,718	2,213
2. Angat	959	568,401	598
3. Penaranda	575	308,965	539
4. Coronel	709	265,787	374
5. Rio Chico	1,177	214,104	182
6. Pantabangan	253	143,353	566
7. Cebu	143	81,158	560
8. Carranglan	258	53,964	209
9. Digmala	52	17,981	345

(B) Cumulative losses from erosion in Upper Agno river basin (million pesos)

Year	Forgone rice production	Forgone power production	Increased amortization	Total
1955	0.00	0.00	0.00	0.00
1960	4.08	1.56	0.59	6.23
1965	15.84	5.88	1.36	29.08
1970	35.40	13.00	2.12	50.52
1975	62.78	22.89	2.89	88.56
1980	97.96	35.58	3.65	137.19
1985	104.96	51.04	4.42	196.42

Sourc: Philippine Environment, 1978.

Range and pasture areas

By law, grazing activities are allowed within the forest lands provided the slope of the area does not exceed 50 per cent. Many leases and permits have been issued without due regard to proper management of the pasture areas. As of 1979 there were 1807 pasture leases and 1826 permits issued covering a total area of 812,512 ha and 154,509 ha respectively. Due to mismanagement of range areas, vast tracts of degraded grasslands with an estimated total of 5 million ha are now common in every part of the country.

Non-timber forest resources

As mentioned earlier the Philippine forests are sources of various non-timber products among which are resins, gums, oils, tannins, ornamental and medicinal plants. During the past, when these resources had only traditional uses, they were harvested indiscriminately. As a result, their volume has greatly declined, and today their availability in the market is limited.

EXPLOITATION AND DEPLETION OF THE FOREST RESOURCES

Statistics on forest resources reveal that some 35 years ago, just after the Second World War, the forests of the Philippines were still intact and many Filipinos believed that they would remain inexhaustible. However, due to the damage inflicted in the various infrastructures by the war, the cutting down of virgin forest had accelerated. Then a few years after the war, the rich, the influential and the powerful became loggers, cutting down trees without due regard for the proper conservation of the forests (Duldulao, 1977). As a consequence, there was a rapid development of the forest-based industries starting in the 1950s up to the late 1960s. This has been attributed to the following factors.

(1) availability of large accessible commercial forest areas;
(2) ease of obtaining cheap war-surplus heavy equipment;
(3) availability of cheap labour;
(4) profitability of log export;
(5) low forest charges levied on timber production;
(6) constant increasing demand for Philippine mahogany logs and other wood products in the international market;
(7) assured local demand for timber and other finished wood products.

The sustained growth of the wood-based industries through the years make them one of the leading sectors contributing to the economy. Many times in the 1960s and early 1970s export earnings ranged from 25 to 33 per cent of the total annual export receipts, making wood products during those years number one export. Hence, from a crude method of animal or steam-engine logging between 1920 and 1940 producing only on average 500 million bd.-ft. of logs annually, and with a relatively low investment, the industries today are highly mechanized and integrated, and heavily capital-intensive, contributing significantly to the economy of the country (Anon, 1974).

MANAGEMENT STRATEGIES AND IMPLICATIONS

We adhere to the basic principles of sustained yield and multiple-use management of our forests.

Selective logging

To maintain the naturally sustainable yield from the uneven structure of the timber stand, a system called selective logging has been prescribed as the regeneration and harvesting system for the Dipterocarp forests since 1954. The silvicultural and management objectives of this method are:

(1) to have adequate number and volume of healthy residuals of desired tree species to ensure future crop and forest cover, for the protection and conservation of soil and water; and
(2) to improve the residual growing stock through silvicultural treatments.

In other words, selective logging is the conservation of the forest by extracting only mature and defective trees and leaving behind enough growing stock for harvest at a predetermined period.

The system involves three phases: namely tree marking, residual inventory and timber stand improvement (TSI). Tree marking ensures that adequate healthy residual stock are left; the inventory of residuals is done after logging to ascertain actual number and condition of the cut-over stand and also for determining punitive action that may be imposed upon the logger. The TSI is a silvicultural treatment designed to remove vines, undesirable trees and other species which compete with the desired crop trees. Initial studies reveal that this treatment can increase the growth of the Dipterocarp residuals two to three times. It is unfortunate, though, that this phase of selective logging has not been implemented extensively except for a few experimental or demonstration areas. The main reason behind is the high cost of this activity, without immediate returns (Pollisco and Natividad, 1980).

Since heavy equipment is being employed in logging the Dipterocarp forests and in the construction of roads, patches of cleared areas are usually left behind after logging operations. This is aggravated by the poor protection measures employed by the loggers in the cut-over area. This situation has encouraged the influx of shifting cultivators (kaingineros) to invade the forest after the destructive loggers have gone. The kaingineros normally followed the path of the loggers and cut the residual trees which were left and burned the forest in order to grow agricultural crops. This practice of shifting cultivation has been going on since time immemorial, and is considered by many observers to be the number one cause of illegal destruction of forests in the country. This is indicated by the result of the joint USAID and Government of the Philippines' study on Mindanao forest resources by comparing aerial photographs taken in 1963 with similar photos taken in 1952. The study revealed that 4.8 million m^3 of timber, or 24 per cent of the total, was due to authorized logging, while 15.3 million m^3 of wood, or 76 per cent of the total loss, resulted from land clearing and other causes (Agaloos, 1965 as cited by Bonita *et al.*, 1977). As a result, during the later 1950s and up to early 1970s the estimate of forest destruction in the Philippines was 172,000 ha per year. At present the partial census of the government on the number of kaingineros in the public forest runs to 154,481 families occupying a total area of 511,645

Table 4 Forest destruction by illegal forest occupants (as of 1979)

Region	Area occupied		Families		Dependents	
	Hectare	Percentage of total	Number	Percentage of total	Number	Percentage of total
TOTAL	*511,645*	*100.00*	*154,481*	*100.00*	*675,714*	*100.00*
1	28,728	6.00	13,769	9.00	64,571	9.00
2	33,337	6.00	13,578	9.00	59,850	9.00
3	19,453	4.00	4,280	3.00	20,399	3.00
NCR	24,366	5.00	5,385	3.00	23,405	3.00
4	17,820	3.00	6,754	4.00	27,862	4.00
5	33,548	7.00	6,034	4.00	32,831	5.00
6	21,004	4.00	4,553	3.00	19,510	3.00
7	47,045	9.00	19,944	13.00	96,613	14.00
8	25,404	5.00	7,700	5.00	41,306	6.00
9	11,461	2.00	2,134	1.00	3,725	1.00
10	134,851	26.00	37,897	25.00	148,437	22.00
11	91,196	18.00	24,047	16.00	112,608	17.00
12	23,432	5.00	8,406	5.00	24,599	4.00

Source: BFD Forest Occupancy and Mangement Section, Forest Protection and Infrastructure Division.
Note: NCR = National Capital Region, which includes Metro Manila.

Table 5 Annual forest destruction in 1979 (ha)

Region	Kaingin and illegal occupancy	Forest fire	Illegal logging	Others	Total
1	1,020.30	3,336.47	150.00	192.00	4,698.77
2	2,440.00	1,973.48	922.00	283.00	5,618.48
3	1,204.06	2,008.00	—	91.00	3,303,06
NCR	2,680.40	1,238.39	98.00	178.00	4,194.79
4	1,016.00	2,330.45	645.00	165.00	4,156.45
5	3,002.00	498.50	450.00	99.00	4,049.50
6	2,841.00	134.20	980.00	345.00	4,300.20
7	3,532.00	2,065.00	—	89.00	5,686.00
8	1,600.00	841.25	1,050.00	238.00	3,729.25
9	2,367.02	227.07	600.00	250.00	3,444.09
10	4,315.98	795.50	1,040.00	418.00	6,569.48
11	2,930.00	1,681.53	1,016.00	331.00	5,958.53
12	3,968.00	1,005.90	1,089.00	425.00	6,487.90
TOTALS	32,916.76	18,135.74	8,040.00	3,104.00	62,196.50

Source: Bureau of Forest Development.
Note: NCR = Capital Region, which includes Metro Manila.

ha, or an average of 3.30 ha per family (Table 4). Based on the yearly census for 1979, the destruction attributed to kaingineros, forest fire, illegal logging and other calamities sums up to 62,196 ha (Table 5). It seems that at this time, during the period of Martial Law in the Philippines, illegal forest destruction has been reduced to a certain degree.

Seed tree method

This is the prescribed regeneration system for the Benguet pine forests. It requires that 16 to 20 seed trees/ha be left to provide for natural regeneration. Initial results reveal that this number of seed trees provides about 570 young trees/ha 10–15 years after logging. This is also the method of regeneration applied to mangrove forest.

Destructive logging in the pine forests had resulted in patches of grassland which are very vulnerable to fire during the summer months when most forest fires occur. Opening up the forest has encouraged the natives to clear-cut the young forest vegetation and convert the area to kaingin-type of farming. One national park within the pine region had been converted by the Ifugao and Igorot natives to an extensive vegetable garden. When confronted by forest officers, they always defy legal orders. Another factor that impairs the productivity of the pine forest is the tapping of the trees for oleoresins. At present there are about 50,000–60,000 tappers within the pine forest. Tappers who are not trained may cause serious injuries to the trees, which may cause their death. Observation reveals that pine trees which have been attacked by *Ips calligraphus* were previously tapped by oleoresin gatherers.

Forest fire also threatens the sustained productivity of the pine forest. This has been a perennial socio-economic problem in the pine region. One study conducted among the Ifugao natives (Codamon, 1979) concluded that the pattern of forest burning among the Ifugaos may be traced back to deeply held beliefs, customs and practices; a sense of awareness and perceptions as to the nature of forests in general which influenced and/or interacted with attitudes held towards forest fires and other forestry-related matters. It was also shown how the government through its implementing arm played a role in the perpetuation of forest burning in the area. In a similar study (Codamon and Atienza, 1980), conducted in the mountains in the vicinity of a large grassland area where burning has been frequent, it was discovered that the tendency to burn a forest appears to be born out of a socio-cultural and political climate whereby the governmental relationship with the local people surfaces as a major issue. According to these authors the kaingineros had a high level of tree-consciousness, yet the majority are not aware of the negative consequences of burning which they accordingly intentionally resort to due to shared beliefs that forest burning would increase soil fertility and enhance the growth of forage for animals. They perceive that forestry laws are ineffective due to non-implementation and consequently show a very low regard for government authorities. The Kaingineros do not want to be involved even in forest fire reporting, partly because they think their efforts will be in vain and partly because they are divided amongst themselves due to varying and conflicting ideologies. They are wary of government-sponsored programmes and the tendency of some is to openly negate what the forestry agency is out to do.

On the inhabitants within the fringes of the mangrove areas, a recent study (Santos, 1980) revealed that most of them are at a subsistence standard of living, having an annual average income of ₱4645 (about US$618) which is very much below the poverty threshold income in the Philippines. Most of the people near the mangrove areas have fuelwood gathering as their major occupation. This practice has resulted to the abuse of the mangrove ecosystem and degradation of the site as a marine habitat. Asked about their goals and aspirations in life, the three most frequently mentioned aspirations were for better and improved life, children's education and sufficiency of family basic needs. Most of them decided to stay in their present residence because of the availability of a source of income in the area, close family ties or loyalty to their birthplace and the peaceful condition in the area.

Legal action

The legal measures that the government has resorted to in order to implement an effective forest management are the traditional punitive measures as a response to law violations. Our experience, however, through the years is that apprehending forest violators did not deter the people from continuing their destructive activities in the forests. Forest destruction goes on with increasing seriousness (Duldulao, 1979). Whenever forest officers insist on doing their job in bringing the case to court, the prosecution has to contend with the following factors: slow phase of court trials, threat of liquidation or death, and most often interference from powerful politicians.

Reforestation and protection

At present the target of massive reforestation work is the 5 million ha of land now grassed that has been denuded of trees, of which 1 million ha are within the critical watersheds which need immediate rehabilitation measures. Prior to 1976 the annual rate of reforestation was only about 20,000 ha against 75,000 ha per annum of forest destruction. In addition, the reforestation problem had been aggravated due to low survival of planting stock so that almost 50 per cent of the yearly reported plantings were actually replantings. Many reforestation areas were burned either intentionally or accidentally by people near the area, so that replanting work could be a yearly source of employment for them.

Alarmed by the critical gap between forest destruction and effective reforestation, the President in 1976 launched a Program for Forest Ecosystem Management (PROFEM) which called for co-operation between the government and private sectors in reforestation. Another Presidential action was the promulgation of PD 1153 in 1977 requiring all citizens 10 years old and above to plant twelve trees a year for five consecutive years. According to the Bureau of Forest Development, these twin moves accelerated reforestation achievement to 79,397 ha in 1979—the second year in history where reforestation has outpaced the reported forest destruction. The success of the reforestation can be attributed to concerted efforts of all sectors with no less than the President spearheading the movement. Protection was intensified by deploying more forest guards in critical areas and deputizing barangay officials, who now total 5667, as forest protection officers.

While it is true that success was gained in reforestation, the greater problem facing effective forest management in the Philippines is the age-old one of upland farmers in the forest land, the kaingineros, who depend for their livelihood on the products they produced from the forests. It has been suggested that the kaingin problem should be viewed in its totality, not only as purely economic in nature but as a combination of economic, social and political problems.

SOCIO-ECONOMIC APPROACH TO FOREST MANAGEMENT

Over the years it has been our experience that technical and legal measures are not sufficient tools in the effective implementation of forest management regulations. In fact in many instances their effect has been negative and has created antagonism between the government and those forest violators who are inside the forest area. Having recognized that forest management could only be effective if socio-economic consideration is built into the system, the government, through the Bureau of Forest Development, has attempted in some pilot areas to implement several developmental projects as socio-economic components of its total forest management scheme. These are as follows.

Kaingin management or forest occupancy system

This approach was conceived by the government as early as 1971 to direct the activities of the kaingineros or the slash-and-burn farmers within the public forest land. It is a key feature of the government forest protection programme with the basic objective of improving the social and economic conditions of the kaingineros by using them as partners in the forest development and conservation programme. A major component of this programme is the continuing census of all kaingineros inside the public forest and the areas they occupy. As of 1979, a partial census shows that there are 154,481 families of slash-and-burn farmers inside the public forests occupying a total area of 511,645 ha. If a hectare of kaingin contains 37 m^3 of timber as initially estimated by the BFD in the 1970s, then the above area must have lost forest trees with a total volume of 18.90 million m^3. Remember that the above report of forest occupancy is only a partial return.

The family approach

This is a reforestation programme started in 1974, aimed at hiring hilly-land occupants and their families to undertake reforestation work to improve their socio-economic conditions (Aguila, 1980). Based on the taungya system developed in Burma, each family was given 5 ha to plant within 2 years, with all the inputs provided by the government. The farmers were paid in five instalments with the provision: that the last payment shall be made when 60 per cent survival of the planting stock has been attained. An evaluation of the programme revealed that the participant farmers have a mean annual income of ₱2513 (US$335).

Communal tree-farms

Established in 1979, this programme of the government aims to augment the income of the upland farmers, and at the same time to complement the efforts of both the government and private sectors in reforesting denuded lands. The programme enables upland communities to obtain long-term leases of 25 years over idle forest land to convert it into tree farms for their benefit. Since this programme is still in its take-off stage, the thrust at this time is to look for suitable areas and co-operators in the various parts of the country.

Industrial tree plantation

In an attempt to complement efforts to accelerate reforestation and to help forestall the impending wood crisis, the government has recently launched a nationwide programme on industrial tree plantations in all open, denuded, brushland and inadequately timbered areas. It is also the social and economic objective of the programme to help uplift the upland farmers. According to estimate, the government would need to develop 400,000 ha of industrial forest plantation by the year 2000 (Revilla and Bonita, 1978).

Industrial tree plantations for pulpwood have been established by big companies, notably Paper Industries Corporation of the Philippines (PICOP) and Nasipit Lumber Company (NALCO) in Eastern Mindanao with farmers as their plantation co-operators. Hand-in-hand with this plantation programme, agroforestry is also a thrust among kaingineros squatting in the concession areas of logging companies.

The latest thrust on dendrothermal plantations has been gaining ground with the farmers in the upland areas.

Lately, the National Electrification Administration (NEA) has contracted with farmers to establish in several areas ipil-ipil (*Leucaena leucocephala*) plantations to supply its dendrothermal power plants.

Food production programme inside timber concession areas

In order to supplement the current food production programme of the government all concessionaires were required to develop areas within their concessions and pasture lands for the production of rice, corn and other basic staples to take care of the consumption food requirements of their workers and the people within the vicinities of their areas who are mostly shifting cultivators. Constraints to the success of this programme are the high cost of production and the rough terrain of the area. As of this time, however, no formal assessment has been done by the government as to how effective these programmes are in relation to their objectives.

REFERENCES

Anon (1974). 'The Potential of Philippine Forest Resources For Industry and Trade Development'. Country introductory lecture, *Symposium on Research and Marketing of Southeast Asian Timbers and Timber Products*, 18–23 November, Philippines.

Aguila, T. L. (1980). Participants' perceptions of some aspects of the family approach to reforestation at Malaybalay, Bukidnon. Unpublished MS thesis, University of the Philippines at Los Baños.

Aruego, J. M. (1973). *The New Philippine Constitution Explained*, 1st edn. University Book Supply, 1117 Espana St, Manila, Philippines.

Bonita, M. L. *et al.* (1977). 'Historical perspective of the Philippine forest resources', paper presented at the workshop on *Natural Resource Environment and the Philippine Forest*, Population Center Foundation, Manila, Philippines.

Bureau of Forest Development (1979). *CTF Annual Accomplishment Report.* Diliman, Quezon City, Philippines.

Bureau of Forest Development (1979). *Philippine Forestry Statistics.*

Codamon, S. L. (1979). Some Socio-Psychological Factors Influencing Forest Destruction by Fire in Ifugao. M.Sc. thesis presented to the Graduate Faculty, University of the Philippines, Los Banos, College, Laguna, Philippines.

Codamon, S. L., and Atienza, P. M. (1980). *A Socio-Psychological Study on the Causes of Forest Destruction by Fire in the MARATAF Area in Pampanga.* A joint project of the Forest Research Institute, College, Laguna and the Pampanga Agricultural College, Magalang, Pampanga, Philippines.

Duldulao, A. (1977). 'Physical characteristics of hill country areas and how to make them contribute to socio-economic development'. Proceedings on *International Workshop on Hilly Land Development.* 3–6 August, Legazpi City, Philippines.

Duldulao, A. (1979). 'Social Aspects of Watershed Resources Management Alternative: Case Analysis'. Paper presented during the *Joint Regional Training*

Course on Watershed Resources Management and Environmental Monitoring in Humid Tropical Ecosystem, held at UPLB, College, Laguna, 5 November.

National Environmental Protection Council: Philippine Environment (1978). *Ministry of Human Settlements*, TRC Bldg., Buendia Ave., Ext., Makati, Metro Manila, Philippines.

Pollisco, F. S., and Natividad, W. R. (1980). '1980 Philippine Forestry Scenario'. Paper presented at the *Workshop on Man, Forest and Environment—a New Forest Policy?*, held at the East–West Center, Honolulu, Hawaii, 3–15 November.

Revilla, A., and Bonita, M. (1978). Quoted by Pollisco and Navitidad (1980).

Santos, M. S. (1980). *Socio-Economic and Cultural Profile of Mangrove Communities in the Philippines*. Forest Research Institute, College, Laguna, Philippines.

Presidents' Center for Special Studies (1980). *The Philippines in Brief.* Office of the President, Republic of the Philippines, Malacañang Manila.

Virtucio, F., and Torres, M. G. (1978). Status of management and utilization of forest resources in the Philippines, *Malaysian Forester*, **41** (2).

Socio-economic Effects and Constraints in Tropical Forest Management
Edited by E. G. Hallsworth

12

Socio-economic effects and constraints in forest management in Kenya

W. J. Lusigi
UNESCO Regional Office for Science and Technology for Africa, PO Box 30592, Nairobi, Kenya

Rapid population growth, limited arable land and shortage of employment opportunities in the industrial and services sectors have resulted into a high pressure on the forest land resources in Kenya. While increased settlement and exploitation of these lands provide a basis for the livelihood of many people, there has been an increase in the deterioration of these areas through over-exploitation and misuse resulting in serious soil erosion. In only about 3 per cent of the land surface does the reliable rainfall exceed the potential evapotranspiration and although seasonal surpluses of water do reach the streams, the areas responsible for supporting vital dry-weather stream flow are very small indeed. When these factors are combined with the traditional farming methods, a situation arises which causes the economy of the rural sector in Kenya to be maintained at subsistence level. But if and when population pressures continue rising and the constant indiscriminate felling of trees and expansion of settlement into remaining forested lands continue, then the resulting environmental damage done may be irreversible.

Recognizing the importance of forests and the dangers of soil erosion resulting from indiscriminate use, the government of Kenya has embarked on an extensive forest conservation programme that has a three-way approach—preservation of indigenous forest, forest plantation programme and rural afforestation programme. The President of Kenya also recently ordered the formation of a Presidential commission on forest conservation and erosion control.

Since Kenya largely depends on the primary production from its land resources, the full potential of the land must be developed, but this must take into account the needs of the local population. Forests do more than just produce. They protect important water catchments and significant genetic resources. Planning must also be based on a realistic appraisal and evaluation of the cultural, political, ecological and socio-economic situation. The approaches adopted must be acceptable to the people and balance resources against local human needs in both the short and long term.

This chapter is not intended to discuss the values of forests for it is assumed that these are well understood. It is intended here to reflect on the serious environmental problems associated with forest use and why they occur; an attempt is also made to discuss possible solutions to the problems and alternative approaches that could be adopted.

HISTORICAL BACKGROUND IN KENYA

The people of Kenya have always been—and still are today—totally dependent for their existence and welfare on a thin, nutrient-filled layer of topsoil, the rain which falls on that soil, and solar energy working through plant life rooted in the soil.

As another vital consideration, Kenya has a severely limited area of really fertile topsoil, and through the impact of one of the highest population growth-rates in the world, a number of (fertile) districts have reached—while some have even exceeded—their land carrying capacity. This is always strictly finite, and beyond its peak level all yields must progressively decline.

At the beginning of this century such a problem was never acute. Many of the lands which are cultivated now were—at that time—heavily forested. The human population was barely more than 2 million (compared with an official forecast of 30 million by the end of the century) and their subsistence requirements were met, using primitive implements, via shifting cultivation with long fallow periods.

Within the 1920s and 1930s, two influences on usage of the national soil complex began to be felt. In the Highlands, a first wave of European settlement resulted in some significant land clearance, plus the introduction of intensive mixed farming, commercial livestock husbandry and plantation crops. Concurrently, in the then so-called African areas, the beginning of an upsurge in human population led to more forest clearance, increasing pressure on soils through shorter fallow periods, some general stabilization of farming practice and mounting fragmentation of the land.

Round about the 1940s colonial development plans applied to agriculture in the 'African areas' embodied investment in a number of technical activities. Of particular relevance here is that urgent forest and soil conservation measures were introduced—covering selected portions of Central, Eastern and Western Provinces—wherever it was most obvious that excessive quantities of soil were being lost from formerly fertile areas. Such measures were strictly enforced, and resulted in some control of the situation, but there was considerable resentment of this entire innovation (in part politically inspired) by the African peoples affected.

During the 1960s, embracing the time of Kenya' independence, two more influences were brought into play. First, and although forest and soil conservation statutes remained legally viable, the emphasis was swung round away from enforcement and towards an advisory approach. In field practice such a change of attitude resulted in substantially reduced concern for forest conservation, especially among new entrants to farming, while many of the terraces and other works completed during earlier decades were allowed to fall apart. Secondly, and against such background, transformation of European farms into African smallholdings or co-operatives characterized one of the world's largest resettlement operations. Something like 900,000 ha eventually changed hands, probably more than half becoming small-scale farms with the balance retained as large-scale units.

Final phases of the major resettlement programme (still sporadically continuing in terms of fresh land units and different techniques) spilled over into the 1970s. At the same time, throughout this latter period, emphasis was shifted to more intensive development of smallholdings in all the traditional African farming areas, which still contain the bulk of Kenya's total agrarian population. Concurrently, there were mounted urgent and widespread projects of land consolidation and registration. Otherwise, these 1970s may be seen as the era of exploding human populations, with no remaining fertile land which could safely be cleared for additional cultivation. In this situation of some trauma, forest and soil conservation measures have largely been neglected. No kind of policy or directive has proved sufficiently meaningful, in its end-product, to outweigh official apathy and human desperation.

EXTENT OF FORESTS IN KENYA

The total forest cover in Kenya, which includes both gazetted and ungazetted forests on large private farms, cover about 3.3 per cent of the country's total land area of 569,250 km^2. When one considers the fact that 88 per cent of the country is arid and semi-arid, and that the bulk of the growing population which is largely agricultural and rural is to be supported from the remaining 8.3 per cent of the country, the picture of the forestry situation for the future does not look very bright. Most of these forests are to be found on mountain slopes of the Kenya Highlands while the rest occur on hillslopes and river valleys on the drier plateau in the hinterland and also along the narrow coastal strip.

The forest types vary with altitude and localized ecological and geological factors. In the highlands where rainfall is high and reliable montane forest types are found where *Ocotea usambarensis, Aningeria adolfi, Podocarpus milanjianus* and *Prunus excelsum africana* dominate the wetter montane areas while *Juniperus procera, Olea hochstetteri* and *Cassipourea malosan* dominate the drier montane slopes. The area above the montane forest is occupied by bamboo forest—mainly *Hagenia abyssinica*. Mixed evergreen and deciduous forests are found on the rest of the highland plateau while the dry upland forest dominates the lower slopes of the rift valley escarpment. Mangrove forests occupy the narrow strip along the coast with a dry coastal forest immediately behind the mangrove forest.

In addition to the indigenous forest cover, the government has embarked on a well-co-ordinated forest plantation programme, so that today over 1400 km^2 of land have been planted, consisting largely of East African cypress, saligna gum, and black wattle.

FORESTRY PRACTICES AND CONSERVATION PROBLEMS

The Kenya Forestry Act lists the following major objectives of the Kenya forest management policy:

(1) the protection of the forest estate to maintain climatic and physical conditions of the country, to conserve and regulate water supplies and to conserve the soil by prevention of desiccation and soil movement caused by water and wind;
(2) the provision of timber and other forest products both for consumption within the country and for export;
(3) the provision of recreational facilities for the public and the preservation of wildlife in so far as is consistent with sound forest management;
(4) provision of employment.

Generally three approaches have been adopted by the forest department to forest conservation.

Protection of indigenous forests

The first approach is to attempt to obtain the complete protection of indigenous forest on water catchments and to some extent in national parks. There has also been some loose protection of indigenous forest in the form of conservation areas, but these do not really have full protection status.

In the montane forest where because of altitude much of it is still out of reach, protection has been reasonably successful so far, but even here the cutting of bamboo for fencing material is commencing. Here it is the question of law enforcement as the activities are all carried out by a few selfish individuals who of course are not ignorant of the law. Outside the montane area the protection of indigenous forests has been largely giving way under human pressures.

The continuing destruction of forest and clearance of bush leads always to greater and more rapid discharge of water, increasing the flood peaks while depressing normal flow between floods. River banks often become eroded and there are so many examples—on a national basis—of dangerously high sediment levels in river systems, indicating by their colour how much valuable topsoil is being swept away.

Ecologists have been advising, for many years, that indigenous forest hacked and burned away for settlement and cultivation will never open up inexhaustible banks of fertility. All that happens is the securing of some temporary (and rapidly diminishing) palliative, in response to political and population pressures, at the cost of wanton damage to the nation's environmental health.

Over recent years the Mau and Mt Elgon forests, as well as forest on the Cheranganis and Chyulus, have been indiscriminately destroyed by squatter-type invading populations, most of whom falsely claim the 'right' of having been born and always lived in such areas.

Kenya's national report of 1972 recorded that—for the Tana River Basin between Seven Forks and Grand Falls—a total sediment yield of 12,300,000 metric tons per annum was estimated. The river at this point drains the Nyambeni hills and the eastern side of Mt Kenya, where steep (and once forested) slopes are cultivated with little or no conservation practice. In parts of Samburu, uncontrolled (and usually illegal) grazing takes place and so as a result of all the burning, trampling and browsing essential forest cover is being destroyed. During recent dry months in Marsabit District and Tharaka area of Meru District, forest grazing may have accommodated 32,000 cattle and 20,000 sheep.

The progress of indigenous forest destruction has in fact become countrywide differing—as between one district and another—only in degrees. The consequence must also (unless this trend is halted and reversed) be countrywide.

Large-scale forest plantation

The Forest Department maintains nurseries in various parts of the country which grow seedlings for the surrounding areas. These nurseries also provide seedlings for the forests owned by the state. Most of the seedlings are of the exotic softwood species, especially *Pinus patula, P. radiata, Eucalyptus saligna* and *Acacia mollissima*.

Large forest estates have so far been perhaps the most successful of the forest conservation exercise. This is easy to understand as they are on state land which is fully protected by the forest department. Most of the commercial timber used in the country comes from these forests and they also feed the only papermill in the country, at Webuye. There is a mounting danger that these forests might not regenerate fast enough to be able to cope with the increased demand from both industry and domestic construction. There does not also seem to be any likelihood that the areas of plantation forest can be increased in view of the increased population. It is largely this realization which caused the government to launch the rural afforestation scheme.

The rural afforestation scheme

The rural afforestation scheme was introduced to mobilize tree planting by the rural masses in order to increase the total land cover by forests. It was introduced at the central government and county council level and at the farmers' level. At the central and local government level relatively little difficulty was experienced since, like the plantation forest, these are introduced on public land and are done at government expense. Promotion of rural afforestation among private farmers, however, encountered some problems. First it was promoted in areas ill-suited for tree growth so that when farmers tried, and survival rate was minimal, they tended to abandon the trial and even developed hostility to the idea. Secondly, with the scarce land resource in the high-potential area farmers were unwilling to divert land available for food to trees which did not generate an equally lucrative product. Fruit trees have been found to be in higher demand than non-fruit trees, which confirms the overriding profit motive among the farmers. Estimates carried out in Machakos District of Kenya show that a hectare under forest gives an average expected economic return that is only about a third of the same piece of land under food-crop production. Prices for wood products, especially charcoal which is widely used by many people, are controlled by government such that the returns from investment in forestry cannot compete with other land-uses. In view of the relatively high costs of alternative fuel, wood will remain the major energy source for a majority of the people in Kenya. In the absence of rural afforestation schemes, this pressure will have to be borne by the few natural forests left and plantation forests.

MAJOR CAUSES OF FOREST DETERIORATION AND DESTRUCTION

It is abundantly clear from the foregoing observations that the problem of forest deterioration is well appreciated by governmental authorities as is shown by the following quote from the national report.

> The conversion of forest land to other uses takes several forms, but since—in all cases—areas being converted are normally climax communities in the form of high forest, there is usually no likelihood of the process being reversed within less than 50–100 years.

The destruction of forests may directly eliminate important water supplies. Decimation of forests is often bringing in its wake costly changes in river regimes.

It is also abundantly clear that the deterioration in the forest resources in Kenya is not due to the fact that nothing is being done. To the contrary, a department exists, plans have been implemented, but the actions have not produced the desired results.

RECOMMENDATIONS

I wish to submit that the reason must still be that the plans implemented seem not to have taken into account the interests and fears of the people of the rural areas who were most affected. Where a population of people lives, there is bound to be also a complex of both socio-economic, political and biological values, and, unless these are incorporated in planning, well-intended programmes will not work.

The problems we have seen in forest management in Kenya are people-oriented. The population has increased considerably and these people are in turn wholly dependent on the soil for their livelihood. In my view no army or laws can keep these people from invading the forests for survival if alternative means to get enough food for themselves are not provided. The first and foremost problem therefore is how to get these people enough food to eat without increasing the land area which in many parts is reaching its carrying capacity. Secondly, the other question to be resolved will have to be that of culture and ways of life. To achieve the required goal in providing enough food for the people might entail a change of ways of life of the people and this will be resisted if planning is not done with their consultation. It might mean, for example, that people who culturally are attached to the land will have to accept work in factories in cities. Thirdly the people will have to be made aware of the consequences of their actions and the problems confronting them if resource deterioration continues. This is not as easy a task as it sounds in view of the fact that we are dealing with people who have fixed ideas on their way of life. This means an acceptable environmental education programme for the masses. This must be made to aim at their wants, e.g. a secure food base for entire families.

Finally the lack of any integrated land-use policy, on a national scale, has been a major constraint. There has never been adequate provision—finance, personnel, equipment—for the fundamental safeguarding of the forest and soil resource, which has not yet been given any adequate dimension in development planning. Such a low-profile attitude must have contributed to the aura of indifference among producers, administrators and local leaders.

New conservation programmes should be urgently addressed to the task of preventing further forest destruction and to any valid possibilities of restoring already deteriorated lands. A fundamental approach must be that of ecological land capability, and all physical conservation measures should immediately be consolidated by policies to step up environmentally sound productivity in both arable and grazing areas.

But really, in its nature and scope, the total environmental threat arising from forest and soil degradation should now be approached and regarded as a national emergency, clearly threatening the physical survival of Kenya.

Socio-economic Effects and Constraints in Tropical Forest Management
Edited by E. G. Hallsworth

13

The forest resource in Kenya in relation to other natural resources

WANJIKU MWAGIRU
Department of Geography, University of Nairobi, Kenya

INTRODUCTION

Forests in Kenya perform two basic and critical roles: a protective ecological role and a productive economic role. Major issues of forest management in Kenya are therefore related to the bio-economic characteristics, functions and interactions of the country's forested land.

More specifically, two sets of macro values can be identified as the more critical variables determining the nature and significance of forest management. Firstly there are the ecological values inherent and present in the forest resource itself. Secondly, there are the economic values superimposed upon the ecological ones by man for purposes of exploitation.

In order to realize these latter values, properties and changes not naturally given in the original formulation of the country's forest ecosystems are introduced and incorporated in the forested zones. Such incorporation induces alterations not only in the forest ecosystems at large but also in the value and functions of forest resource products (Muiruri, 1974, 1977a).

At another level, Kenya's current level and pace of development precipitates other social issues more directly emanating from the human population and which ultimately also have repercussions on the management of the country's forest estate. Urgent socio-economic demands upon the forest are related to a number of factors that are symptomatic of the type of development problems facing the country. Some of these include the scarcity of high-potential agricultural land relative to the size of the population; rural and urban rquirements for both domestic and industrial uses of water, as well as demands related to timber for the construction industry and pulp for paper manufacture. In addition, there is the even more urgent demand for fuel, firewood and charcoal for household energy requirements.

It is therefore virtually impossible to ignore the direct impact of man upon the forested environment in Kenya.

An examination of the ecological and economic aspects of forest management in the country should therefore aim at inspiring a wiser use and conservation of this resource. At whatever level of magnitude the exploitation of the Kenyan forests is undertaken, the overriding principle should be one of conserving the inherent bio-economic productivity of the forest ecosystems and their related resources of soil, water and wildlife. A strategy of such 'limited' conservation could have one or both of two goals:

(1) not to push forest resource exploitation further than a point of adequate yield;
(2) to try and preserve the basic structures and functions of the exploited ecosystem.

The following discussion will examine the macroscopic properties and functions of the Kenyan forest resource and, in view of these, present an analysis of its management implications. The underlying theme is the search for the essential criteria to be used in assessing and controlling the process of exploitation and the result of the conservation measures undertaken. Throughout this discussion, the basic assumptions are: firstly, that value conflicts arise when a commodity-oriented economy develops the desire to restore and maintain the quality of its natural and human environments; secondly, that the general requirements for the development of an effective management strategy for the Kenyan forest ecosystems depend upon a proper, rational conceptual base (Muiruri, 1977a).

THE FORESTED LAND IN KENYA: AN OVERVIEW

Kenya has a total land area of 569,250 km^2 and of this, only 12 per cent is arable under existing levels of agricultural technology. This is the only part of the country that is agriculturally productive—a belt of high-potential land running roughly south-east of Nairobi to Lake Victoria, plus the narrow coastal strip. These are the areas that receive adequate rainfall for intensive crop farming. These are also the areas of the greatest population concentration. Surface water availability, mainly in the form of rainfall, is the most critical factor in determining land productivity, and therefore population distribution.

Land in Kenya has been defined in various ways but nearly always in terms of the natural resources within it or its general land-use capability. On the basis of broad land potential categories, land has been classified as shown in

Table 1

Total land area	56,925,000 ha
Water area	1,123,000 ha
High-potential land	6,785,000 ha
Medium-potential land	3,157,000 ha
Low-potential land	42,105,000 ha
Other	4,867,000 ha

Table 1. About three-quarters of the total land area comprises the low-potential arid and semi-arid areas which are relatively low-yielding and which require special technological and management inputs in order to become productive. The other high- and medium-potential lands support the bulk of agricultural and other land-use activities such as forestry.

Kenya's forests cover about 3 per cent of the total land area and about 25 per cent of the high-potential agricultural land. This forested environment is generally considered as being critically small and significant attempts have been made in the past, and are still being made, to increase the area under forests, to conserve it, and to exploit and manage it efficiently and judiciously.

Firstly, it is on the spatially restricted high-potential land (highland and montane areas) that the greatest proportion of Kenya's forest, water and some wildlife resources are located.

The country's forest estate contains some of the most important natural resources and its well-being directly influences the climate, soil, water and wildlife resources in the country. All the major rivers of Kenya have their sources in the forested mountain and highland drainage catchment basins.

Secondly, land provides the only direct means of livelihood for 87.5 per cent of Kenya's 16 million people. The majority of these people are concentrated in the spatially restricted better-watered highland areas of the country. When we relate the socio-economic demands of this population to the proportion of the country's productive high and medium potential land, the problem emerges (Table 2). The country's annual population growth rate becomes critical therefore in view of the spatially limited arable land, and the dominance of the agricultural sector in the national economy. A realistic perspective of the land pressure situation in relation to the future status of the forest estate is better obtained by considering four variables:

(1) the land's potential agricultural productivity and its ability to support people;
(2) the location and size of the forested lands;
(3) the location and characteristics of the major land-use systems;
(4) the management and land-use policies adopted for various key eco-regions and natural resources.

Table 2 Projected per capita land availability (in ha) in the high- and medium-potential areas of Kenya on the basis of total population and the labour force

Year	Total population ('000)	Land availability (ha per capita)	Labour force ('000)	Land availability (ha per capita)
1980	16,053	0.62	5,517	1.80
1985	17,310	0.51	6,658	1.49
1990	23.302	0.43	7,900	1.26
1995	28,213	0.35	9,537	1.04
2000	34,286	0.29	11,553	0.86

Source: Muiruri, 1977.

The paucity of alternative labour-intensive industries outside agriculture will inevitably lead to an increasing shortage of agricultural land with adequate rainfall, which can be managed economically on a smallholder basis under existing levels of agricultural technology. Such land pressure will almost certainly be exerted on the already spatially restricted and congested arable lands. It has been stated that for Kenya's rural population (87.5 per cent of the total), the crucial problem is not the availability of wage-earning jobs *per se* but the availability of land plus the knowledge and supporting services to farm it well and obtain a reasonable income (Ominde, 1968).

The man–land ratio therefore has grave implications for the continued bio-economic productivity of the vital high-potential land. Due to continuous pressure on the limited available land, more and more people in the country have tended to see the forested land in terms of its potential for crop farming. This dimension of the land problem has serious implications for the continued stability of the economy–environment interaction in Kenya *vis-à-vis* the forest resource. Excision of gazetted forested land for the creation of land settlement schemes has been done in the past in order to ease socio-economic and political pressure, and it appears that such pressure is likely to intensify. The problem is therefore one of relating the need to 'preserve in perpetuity' the limited forest resource, to its much wider socio-economic framework and to the existing policy and legislation.

A fundamental problem of forested land management and conservation is related to the existence of conflicting and competing policy objectives adopted by various government institutions for the utilization of the lands carrying important natural resources. The problem has aptly been summarized thus:

In Kenya today, we have reached a situation where land-use interests such as agriculture, tourism, ranching, wildlife management, forestry and water conservation—each one of them valid and nationally productive uses of land—are in some instances in competition and often in conflict over large areas of the country. Not only are the various arms of Government in disagreement or confusion on these issues but this is compounded by the demands of the landless and burgeoning population which are haphazardly realized in the absence of clear policies (Government of Kenya, 1972).

The utilization, management and maintenance of the major productive eco-regions of Kenya are highly compartmentalized and so are the policies formulated by the various government departments for such utilization and management. This compartmentalization brings about conflicts at the most important levels of policy- and decision-making. The challenging problem is one of achieving the diverse objectives of these institutions in a way that does not minimize but enhances the productivity of the natural resources within the various eco-regions (Muiruri, 1977b).

FORESTS AND THE WATER RESOURCE

Relative to area and population, Kenya has extremely limited surface and groundwater resources. For example, only 5 per cent of the total annual precipitation goes into surface run-off, and this almost wholly originates from the forested drainage catchment basins. Its effectiveness is further reduced by intense evapotranspiration. Perennial rivers are largely concentrated in the lake basin, the central and coastal areas of the country, while 2 per cent (13,396 km^2) of the total area of the country consists of water bodies.

More than any other resource sub-system, the water factor links the high-potential eco-regions with the national socio-economic system. In Kenya, land-use problems are inseparable from the linking problems of water-use and water resources. From an economic viewpoint, rural and urban water investments are an essential part of the socio-economic infrastructure, whose needs and priorities cannot be adequately evaluated by reference to any one criterion, be it economic, ecological, financial or social.

Kenya's water development strategy is to provide adequate running water to every household by the year 2000. To this end, investigations are currently under way to provide data on:

(1) availability, reliability, and quality of the country's water resources;
(2) a forecast of demands over the next 10, 20 and 30 years.

In all the related developments, the most crucial factor is to ascertain long-term availability of water for the various uses. This brings in the question of the land-use, management and conservation practices in the major forested water catchments.

The prerequisite for rational catchment management has been said to be, first and foremost, a determination of the hydrological requirements from the catchment, and secondly to decide whether these are best met by increasing the amount of water which infiltrates into the soil.

For Kenya, since surface water storage facilities are limited, it is important to obtain maximum infiltration into the soil, to maintain the yields of the streams throughout the year. The main problems therefore lie in storage, irrigation requirements and watershed management. These three factors, unless well balanced and handled, have the potential for completely upsetting the current water balance.

The stream flow of rivers emanating from the forested catchments reaches a peak during the heavy rains of March and May, and averages much lower flow for the remaining months of the year. That any flow occurs during dry seasons reflects the function of the forested watersheds. Under rational and efficient management, the deep volcanic forest soils facilitate free percolation of water beyond the root range of most trees, then they slowly release the water throughout the year via springs. Forests are said to transpire more than any other type of vegetation, and thus only where rainfall is high can stream flow be expected to continue throughout the year (McCulloch and Dagg, 1965).

Where precipitation rate does not exceed evapotranspiration rate, no water yield may be expected, and the stream curve would return to zero after the wet season floods subside. Such conditions are likely to prevail after afforestation. Because of the lack of vegetation cover or its ineffectiveness, there would be less infiltration and catchment during the rains and this would raise the peak of stream flow at this time.

Flooding and erosion would result, and subsequently there would be zero flow during the dry season. Storage of the peak flow water volume is therefore necessary in order to ensure annual water availability (Muiruri, 1977a, b).

The most pressing question at present is how much alteration the forested watersheds can take and still return a significant flow throughout the year to meet increased domestic, agricultural and industrial water demands. This question has received much attention over the last 26 years and answers are being sought from continuous watershed experiments (Pereira, 1954; Parry, 1962).

The land-use debate has been centred on the water balance implications of large-scale replacement of an indigenous forest cover with either exotic forest plantation species (pines and cypresses), arable crops or tree plantation crops such as tea. In this respect, the various workers have been concerned with establishing the bearing on the water resource of large-scale land-use changes.

The major forested catchments, especially in Central Kenya, have an unusually high potential for the development of softwood plantations. Yet it is also acknowledged that the climate of these areas has a highly seasonal rainfall regime, and in the absence of major water-storage construction works, the farming communities in the surrounding agricultural areas are critically dependent upon dry-season flow. The forest management problem therefore has two dimensions:

(1) a general problem of the hydrological effects of large-scale rotational planting, tending and felling of softwoods in bamboo-forested stream-source areas;
(2) a local problem of the effects of this change in land-use on the river systems supplying water to rural and urban communities.

The presence of forest cover is important in lessening the contrast between dry and wet seasons' flow. The major concern therefore is on the effect of land-use changes on the water economy of the forested drainage basins. Results from ongoing land-use experiments have shown that the role of the indigenous forest cover is particularly crucial in relation to the soil, evapotranspiration rates and water-use, by the different associations of both indigenous and exotic forest vegetation.

FORESTS AND THE SOIL RESOURCE

The soils of the forested catchments regulate both surface and underground waters of the drainage basins, as well as determining the characteristics of the vegetation cover upon which many other habitat components depend.

Their status and condition, relative to changes in land-use practices, have been investigated as part of land-use hydrology experiments conducted in the East African forested water catchments over a number of years (Pereira, 1954, 1963; Robinson, 1967).

Results seem to indicate that the soil–water regime is not adversely affected by changing the indigenous forest cover to exotic forest plantations. However, it has been shown that despite the deep volcanic soil profile with high infiltration rates, the effective groundwater storage capacity is only 76 cm. As a result, the ratio of monthly total wet-season flow to dry-season flow can be as high as 20:1. There is thus a need to construct major storage structures in order to achieve a better stream flow distribution over the year.

Experiments to determine the effect of softwood exotic plantations on soil fertility have not been as positive, however. Robinson (1967) has observed that replacing the indigenous forest cover with exotic softwood species results in an overall decline in soil fertility. Apparently, the 25–30-year rotation period of the softwood forest, though permitting some level of nutrient build-up in the soil, does not facilitate the complete recovery of the nutrient cycle equilibrium that existed beneath the indigenous forest cover.

Nevertheless, a study conducted on one of the forested catchments has emphasized the role of the soil in regulating the water balance of the forested drainage catchment zones. The study concluded that as a natural long-term medium of water storage, the soil is unsurpassed by artificial storage structures. The critical factor of the soil–water regime arises from the seasonality of precipitation. Stream flow becomes reduced during the dry season and this affects the volume of water reaching the lower-lying agricultural and urban areas (Muiruri, 1977a).

FORESTS AND THE WILDLIFE RESOURCE

The forested environment in Kenya is an important reservoir of the wildlife resource. There are three mountain national parks, viz.: Aberdare, Mt Kenya and Mt Elgon national parks.

Game animals interact mainly with the indigenous forest for food, shelter and escape cover. Interaction with the exotic plantations, however, is nearly always characterized by conflicts through game damage to the tree crops.

The type and degree of game–forest interaction in the mountain national parks emphasizes the need for scientific management of the wildlife populations. Detailed ecological surveys to determine the status of the animals in the forested national parks cannot be undertaken without reference to the forest estate, since management of the wildlife resource is wholly tied up with the activities of the forestry sector.

Investigations of game–forest interactions and conflicts, and their implications for game policy and management, have been carried out over various parts of the forest estate in Kenya. Different workers have emphasized different aspects of the problem, e.g. damage caused by big herbivores such as elephant, buffalo and rhino; damage caused by monkeys; the nature and type of damage, that is whether primary or secondary; and finally, the more comprehensive question of the implication of game damage to policy, legislation, management and perceptual aspects of other resource subsystems (Muiruri, 1974, 1977a).

In all these studies most workers have tended to centre their investigations on the question of whether or not big game are to be regarded as a forest pest or a forest resource. Of greater relevance, however, is the more fundamental issue of how to institute a rational integrated management and exploitation strategy that recognizes the ecological interrelationship, interactions and dependencies amongst all the resources of the highland and mountain water catchment and forest ecosystems. Such an integrated approach would at the same time maximize the economic output of such resources.

PROPOSALS FOR FOREST MANAGEMENT

In this chapter the various factors that must constitute the information sources for a management and exploitation strategy of the Kenya forest system have been presented and discussed in detail. These range from ecological, economic, social and administrative factors, to the potential and resultant constraints on productivity largely through man's activities and lack of co-ordination.

The emphasis has been that rational resource management at the ecological and socio-economic levels (the ecosystem–society environment relationship) is a long-term goal, and the progressive introduction of systemic approaches for better human decision-making should be adopted for Kenya's various eco-regions. But the problem is two-fold as the borrowed quotations below adequately summarize:

> Neither in politics nor in administration has there been a generally accepted body of knowledge or doctrine by which conflicts over resource uses could be readily resolved. In the absence of an 'ordering' or organising concept, efforts to coordinate natural resources policies have been largely ineffectual or have been used as covers to impose or to prevent one use over others (Edel, 1973).

> Land is a substantially inelastic resource and this means that as human population multiplies, land policy is increasingly determined by population policy. The inseparable connections between land use, population and public interest are now evident (Ketchum Bostwick, 1972).

No adequate and comprehensive land-use plan exists as the basis of utilization and management of any of Kenya's resource zones. Yet such a land-use strategy is urgently necessary in order to be able to handle the enormous variety of the ecological, economic, social and environmental data available.

The recommendations forwarded in this discussion are:

First, that the entire process of such an integrated land-use approach must be conceptualized in relation to components, processes, resources and use of the resource ecosystems. An integrated, comprehensive land-use plan should be judged on the basis of feasibility, reliability, economic utility and relative freedom from mechanical and administrative constraints. In its progressive development the general design must be constantly monitored by competent judges from a multidisciplinary 'bench' for its intuitive and practical validity. In other words, the plan must represent and reflect the reality of the wider socio-economic environment.

Second, Regional Management Authorities should be created for each resource ecosystem. Such authorities would provide a central co-ordinating point for the input and output factors of the system management process. These bodies should consist of personnel derived from all the departments and bodies involved in the use and maintenance of the resource ecosystems and their resource sub-systems. Each government ministry would provide inputs via its various departments. Such inputs would consist of types of information dealing with ecological, economic, technical, financial, administrative and legal aspects of the resource ecosystems' utilization and management. Local or regional representations would come at the provincial and district levels.

Such Management Authorities would be relatively independent bodies but the relevance and efficiency of their functions would depend on the level and types of inputs from the administrative and management phases. They would also act in a supervisory capacity and in this respect would work closely with monitoring systems of the management and the use-impact phases.

The proposal to create such bodies arises from the need to expand control of the vital eco-region and ecosystem matters in order to reduce central control of compartmentalized ministries and departments as is now the case. Such national/regional control of the eco-regions would seem to be a more acceptable compromise between the highly compartmentalized central control which is apparently not responsive to comprehensive and local system needs, and purely local control which may be inadequate to cope with the scale of the resultant management problems.

Each resource-using body e.g. Agriculture, Game, Forest, Water, Settlement Departments, should be required to prepare and present 'impact statements' to the Management Authorities for any proposed major land-use changes within and in the vicinity of the eco-regions. These impact statements would be submitted to the Management Authorities prior to the implementations of such land-use changes.

The Management Authorities would, on the basis of this and other information, either accept or reject such proposals, once they determine whether or not they would have negative impacts on the total system or on aspects of it. Deliberate changes in the way the resources of the eco-regions are used must be made if negative environmental impacts are to be avoided. Decisions about the use of any resource unit should therefore not be made in isolation from the broader environment system of which it is a part. In the final analysis, the future of the resource zones ultimately depends on man's ability to control the impact of his activities on the resource sub-systems.

Thirdly, concurrent with the creation of the Management Authorities should be the formation of a multidisciplinary resource ecosystem 'task force' to formulate the management programme. By implication, there must also be concurrent development of legal institutions and procedures to make the management of the eco-regions and ecosystems more effective.

Fourthly, a regional and national monitoring system should be formed to continually collect chemical, physical, biological and socio-economic data that have a bearing on the dynamics and stability of the eco-regions and their resource units.

Such a monitoring system should be in a position to give advanced warning on conditions that could be hazardous to the ecosystems. From such information it whould be possible to develop a model for evaluating the potential effects of conceived alternative courses of managerial action. It would ultimately be the responsibility of Management Authorities to determine and generate the necessary kinds of information essential to the development, evaluation and application of the model.

What is being called for is therefore the development of a rational managerial, utilization and administrative framework within which the bio-economic resource attributes may be successfully regulated.

Fifthly and finally, there must be an acceleration and expansion of research and surveys of the environment–economic interface in Kenya, particularly inventories of all the resource sub-systems, including an evaluation of demographic patterns of the human population. In other words, an up-to-date compilation of baseline ecological and socio-economic data of the natural and modified eco-regions in the country.

REFERENCES

Edel, M. (1973). *Economics and Environment*, Prentice Hall, New York.

Government of Kenya (1968). *A Forest Policy for Kenya; Sessional Paper No. 14.* Government Printer, Nairobi.

Government of Kenya (1972). *Kenya's National Report to the United Nations' Conference on the Human Environment.* Stockholm.

Government of Kenya, WHO (1973). 'Surface water resources in Kenya'. Report No. 16, *Sectorial Study and National Programming for Community and Rural Water Supply, Sewage and Water Pollution Control*, Brazzaville.

Government of Kenya (1979). *Kenya National Development Plan, 1979/1983.* Government Printer, Nairobi.

Bostwick, H. (ed.) (1972). *The Water's Edge: Critical Problems of the Coastal Zone.* M.I.T. Press.

McCulloch, J. S. G., and Dagg, M. (1965). Hydrological aspects of protection forestry in East Africa. *E. Afr. Agric. For. J.*, **XXX** (4), 390–4.

Muiruri, W. (Mwagiru) (1974). The Aberdare ecosystem. *J. East. Afr. Res. Dev.*, **4** (1).

Muiruri, W. (Mwagiru) (1977a). *The Aberdare Ecosystem: Its Bi-economic Functions, Interactions and Problems of Management*. Ph.D. Thesis, University of Nairobi, Kenya.

Muiruri, W. (Mwagiru) (1977b). 'Land use on Mount Kenya and the surrounding areas'. Report prepared for the G. of K./WMO/UNEP, the Advisory Working Group on Mt Kenya Baseline Monitoring Station.

Ominde, S. H. (1968). *Land and Population Movements in Kenya.* Heinemann, London.

Parry, M. S. (1962). Progress in protection of streamsource areas in Tanganyika. *E. Afr. Agric. For. J.*, special issue, pp. 104–6.

Pereira, H. C. (1954). The physical importance of forest cover in the East African Highlands. *E. Afr. For. J.*, **19** 233–6.

Pereira, H. C. (1962). Hydrological effects of changes in land use of softwood plantation and Bamboo forest. *J. Soil Sci.*, **B** 299–313.

Robinson, J. B. D. (1967). The effects of softwood crops in the chemical fertility of a tropical soil. *E. Afr. Agric. For. J.*, **23** 175–89.

Socio-economic Effects and Constraints in Tropical Forest Management
Edited by E. G. Hallsworth

14

Amazon forest management for wood production: an assessment of limitations and potentials based on field studies at San Carlos de Río Negro, Venezuela

CHRISTOPHER UHL, CARL F. JORDAN
Institute of Ecology, University of Georgia, Athens, Georgia 30602

and

RAFAEL HERRERA
Instituto Venezolano de Investigaciones Cientificas, Centro de Ecologia, Caracas 101, Venezuela

INTRODUCTION

The Amazon forest has been used relatively little for commercial lumber production compared to other areas of the tropics (Grainger, 1980). There is an increasing demand, however, to put Amazonian forest to use for wood commodities. It is reasoned that because Amazonia is forested, the growing of trees for lumber and pulp is a practical land-use alternative.

The purpose of this presentation is to explain why nutrients are a major constraint to timber production in the Amazon Basin and to suggest, based on field studies at San Carlos de Río Negro, Venezuela, some methods of timber production which conserve ecosystem nutrient stocks.

NUTRIENT LIMITATION TO TIMBER PRODUCTION IN THE AMAZON BASIN

The Amazon Basin, with a land area of about 7.6 million km^2, occupies more than 35 per cent of South America. While there are many different ecosystem types present within Amazonia (e.g. mangrove forest, montane forest, savanna, etc.), about 85 per cent of the basin contains forest on non-flooded ground (Prance, 1978). Of course, many subdivisions exist within this non-flooded forest category (e.g. high forest with closed canopy, low forest with reduced biomass, liana forest, dry forest, etc.), but most of the non-flooded forest is of the high forest type (Prance, 1978). There are also many soil types present in Amazonia (Wambeke, 1978). Soils may be shallow or deep, sandy or clayey, high or low in organic matter, etc. However, the wide array of Amazonian soils have in common that they are heavily leached, acidic, and generally low in mineral nutrients (Beek and Bramao, 1968; Wambeke, 1978). Hence, as a generalization, most of Amazonia contains high forest on severely leached, acid soils. Indeed, much of the forest in the vicinity of San Carlos de Río Negro, Amazon Territory, Venezuela, where we have conducted our studies, fits this description.

The terrain of the San Carlos region consists of a system of rolling hills, some rising as much as 50 m above the Río Negro mean river level (Herrera, 1977). Soils, topography, and vegetation are closely matched (Figure 1). The higher areas are covered by a species-rich evergreen forest 25–30 m tall (Uhl and Murphy, 1981a). The soil here consists of a sandy A horizon overlying kaolinite clay and lateritic concretions. With the exception of the sand cover, these soils are similar to the Xanthic ferralsol type which is mapped as the dominant soil in the central Amazon Basin (FAO–UNESCO, 1971) and are equivalent to Ultisols under the US soil classification system. The lower areas

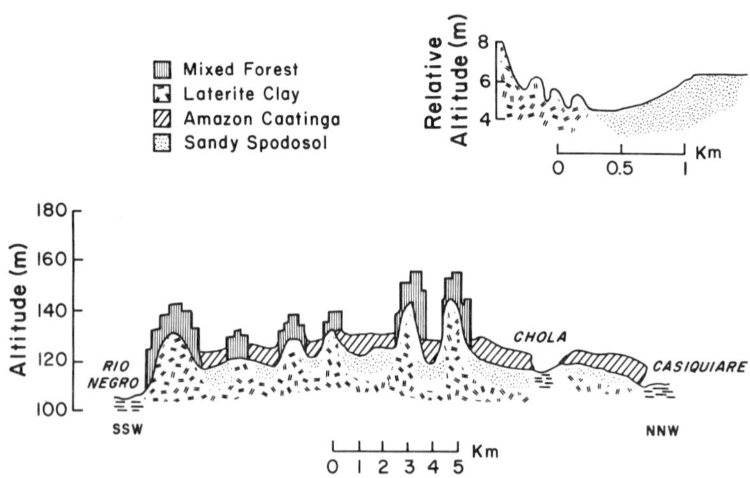

Figure 1 An altimetric transect between the Río Negro and the Casiquiare river showing maximum river level for 1976. The inset at the upper right shows the microrelief in the lowland area

between the hills have seasonally waterlogged soils composed of a thick humus layer overlying highly leached coarse sand and are classified as Spodosols (Klinge and Herrera, 1978). These soils support Amazon caatinga forest, a closed sclerophyllous forest 20–25 m tall showing strong species dominance. Both Ultisols and Spodosols are very low in nutrients because of intensive leaching under humid tropical conditions for millions of years and the absence of unweathered parent material which could serve as a fresh source of nutrients (Fittkau *et al.*, 1975).

The Amazon forest ecosystem, as a whole, differs markedly from many other tropical forest ecosystems in that nutrient stocks are comparatively low, especially in the soils. Nutrient paucity is reflected in the waters of Amazonia's streams which are among the poorest in the world in nutrient concentration (Furch, 1976). Furch and Klinge (1978), studying nutrient concentrations of Amazon vegetation, soils, and stream waters, found that the nutrient content of the waters directly reflected the nutrient status of the ecosystems from which the waters flowed. Herrera *et al.* (1978a), in a comparison of nutrient stocks and concentrations among selected temperate and tropical forest ecosystems, concluded that Amazonia is somewhat below the world average for nutrient stocks in vegetation and nutrient concentrations in drainage water, and much below the world average for nutrient concentrations in soils. Apparently, as an adaptation to the low cation exchange capacity of the soils, cations are stored predominantly in the organic matter (living and dead) compartment of the ecosystem. For example, more than 90 per cent of the total stocks of K, Ca, and Mg in the San Carlos Ultisol forest are in organic matter (Herrera *et al.*, 1981). Similar patterns hold for the caatinga forest at San Carlos (Herrera, 1979).

The low nutrient content of Amazon forest soils brings up the question of how this forest is structured to ensure both a minimal loss and an optimal capture of nutrients. We have identified a series of characteristics associated with the undisturbed forest at San Carlos which either prevent nutrient loss or facilitate nutrient capture. The most striking of these characteristics is a soil surface network of fine roots intermixed with organic matter in various stages of decomposition. In Central Amazonia (i.e. the most nutrient-poor region of Amazonia) this surface root mat can reach a thickness of 30 cm. As one travels away from Central Amazonia (e.g. towards the Guiana Shield to the north) the root mat becomes thinner.

Direct observation of the surface root mat gives some idea of its role in nutrient conservation. When leaves, fruit, or branches fall, roots grow upwards (i.e. in a negative geotropic fashion) and soon cover and attach to the litter. A few months later all that is left is a shell of roots, in the shape of the decomposed litter.

Direct transfer of nutrients from litter to roots in regions of low soil fertility was suggested by Richards (1952). Mycorrhizal fungi which grow between decomposing litter and roots may mediate this direct transfer (Went and Stark, 1968). We have observed mycorrhizal threads in our study site by gently pulling fine roots away from the litter to which they were attached. Herrera *et al.* (1978b) demonstrated movement of radioactive nutrients from litter to living roots through fungal hyphae at San Carlos.

Mycorrhizal fungi, however, are not the only nutrient-conserving mechanism of the surface root and organic matter layer in Amazonian forests. Stark and Jordan (1978) have shown that when a radioactive solution of phosphorus and calcium is sprinkled on top of the root mat, less than 0.1 per cent of the activity is lost from the system by deep leaching, the rest being taken up and eventually translocated to nearby trees. Apparently the immediate retention mechanism is surface absorption on organic material, and later, gradual uptake by living roots occurs. These results, together with the fact that up to 60 per cent of the biomass of some Amazonian forest occurs in the roots, is evidence that the superficial rooting habit constitutes an important nutrient-conserving mechanism because it prevents nutrient movement into the mineral soil where leaching losses will occur.

Epiphyllous organisms, such as mosses, algae, lichens, and bacteria also play an important role in nutrient conservation. Some of these fix nitrogen from the air. Presumably this nitrogen becomes available to the trees when the leaves are shed and roots invade the litter. These organisms also may play an important role in scavenging nutrients from the rainfall.

The characteristics of the leaves themselves, also favour an efficient use of nutrients by their long active lives, their retranslocation of mobile nutrients before leaf-shedding, and their high polyphenol content and coriaceous nature which reduces herbivory while leaves are physiologically active.

The effect of the various nutrient-conserving mechanisms is that the Amazon forest, along with its associated micro-organisms, acts as a giant ion-exchange column that extracts nutrients from water passing through the ecosystem. The net result is that concentrations of many nutrients in the stream flow that drains the ecosystem are lower than the concentrations in the rainfall entering the system.

The common factor about these nutrient-conserving mechanisms is that they are all part of the living, natural forest, and they can be destroyed when the forest is removed, leaving the land in an extremely unproductive state. This was dramatized by the bulldozer clearing of several hectares of well-developed caatinga forest at San Carlos. After bulldozing, the site was left to natural succession. Three years later less than 50 per cent of the ground surface was covered by vegetation. Above ground biomass was a mere 77 g/m^2; 97 per cent of this biomass was composed of forbs and grasses (Uhl *et al.*, 1981a). The few trees that had established on the site were less than 1 m tall and appeared to be under severe nutrient stress. If biomass continues to accumulate at its current rate on this site, more than 1000 years will be required to reach levels of above-ground biomass characteristic of the mature forest (i.e. about 300 metric tons/ha). Sanchez (1976) also reported that mechanical clearing destroyed the productive capacity of the land in the Amazon region of Peru. Hence, any land-use alternative for the Amazon must consider the central problems of nutrient retention and nutrient replenishment.

POSSIBLE WAYS OF MANAGING AMAZON FOREST LANDS FOR TIMBER PRODUCTION

There are three general ways to manage Amazonian forest for timber production:

(1) selective harvest of natural stands preceded or followed by various strategies to increase the abundance of desired species;
(2) clear-cutting and wood harvest followed by natural regeneration; and
(3) clear-cutting and harvest followed by the establishment of tree plantations.

Although our investigations at San Carlos have not been designed to directly assess the relative merits of these different forest-utilization plans, the long-term forest dynamics and productivity studies, and the studies of natural regeneration following different methods of land clearing at San Carlos, do provide some insights into forest management for wood production.

Selective harvest of natural stands

The great diversity of tree species in the Amazon forest, coupled with the lack of an international market for many of the wood types, have made selective logging common in the more accessible regions of the Amazon (Palmer, 1977). Most harvesting is done by river-dwellers acting in conjunction with local agents (Grainger, 1980).

Actually, only about 50 of Amazonia's many thousands of species are widely exploited (National Academy of Sciences, 1980). Hence, when a hectare of forest is harvested selectively, only a few trees are removed. Provided these trees are carefully extracted from the site, nutrient losses are minimal (i.e. only a small percentage of the total nutrient stocks is lost and these losses can be replaced by precipitation inputs), and overall disturbance is not much greater than what the forest experiences due to natural tree fall. However, selective harvesting operations are not always careful. Surveys in South-east Asia reveal that between one-third and two-thirds of the trees remaining after selective logging are damaged beyond recovery (Kartawinata, 1975). Furthermore, as much as one-third of the ground may be left bare after logging (Burgess, 1973).

In the Amazon forest there are many seedlings and saplings on the forest floor. This advance regeneration can persist for long periods of time until tree removal or natural tree-fall openings favour its growth (Uhl and Murphy, 1981a). Because of the importance of existing regeneration in forest recovery, particular care must be taken to reduce damage to this component when large trees are extracted in commercial forestry operations. Clearly, forests such as the Amazon forest, with its low nutrient stocks and soils with low nutrient retention capacity, will suffer greatly from careless harvest operations.

Given the need to conduct tree harvest operations carefully, and the fact that tree seedlings and saplings are abundant in the understory, shelterwood forestry, such as that practised in Nigeria (Lawton, 1978), might have application in Amazonia. The aim of the shelterwood system is to produce a large number of saplings of the desired species before selectively harvesting canopy species. Initial surveys are necessary to locate promising understory individuals and to clear competing vegetation away. Once the desired species are well established in the understory, the valuable canopy individuals are carefully harvested, and ideally, the previously favoured saplings and seedlings then dominate in the regeneration. This is obviously a labour-intensive approach, but in forests where valuable tree species are present it may represent a viable management option. Where no effort is made to influence regeneration following selective harvest operations, desirable species may be permanently eliminated from the forest.

Hence, studies of forest dynamics at San Carlos suggest that to maintain desirable species in Amazonian forests under selective harvest management, forest planners should:

(1) identify seedlings and saplings of desirable species before harvest and favour these through weeding; and
(2) avoid damage to saplings, seedlings, and topsoil in the timber-extraction process.

If these precautions are followed, the nutrient-conserving mechanisms characteristic of the native forest will remain intact and sustained-yield forestry will be possible. This approach will become increasingly attractive, in an economic sense, as tropical hardwood reserves in other regions are exhausted and the value of Amazon timber rises.

Clear-cutting and tree harvest followed by natural regeneration

The obvious disadvantage to selective harvest forestry, in an economic sense, is that only a small fraction of the forest wood production is used. Now that the technology is being developed to handle species-rich tropical hardwood forests (Kyrklund, 1976; Collardet, 1976; Stager Carvajal, 1978), forestry options based on clear-cutting and natural regeneration merit attention. In particular, prospects are becoming increasingly favourable for using multi-species stands for the production of wood chips and fibres (Erfurth, 1976). The principal danger in Amazon forest clear-cutting is that a major portion of the nutrients will be lost from the site via harvest, leaching, and erosion. At San Carlos we have studied natural regeneration on sites which were clear-cut and immediately abandoned and sites which were clear-cut, burned, and then abandoned. These studies illustrate the consequences of different forest clear-cutting management schemes for the Amazon.

Immediately abandoned clear-cuts

The most striking feature of regeneration on immediately abandoned clear-cut sites is the abundance of trees and vine sprouts arising from cut stems or roots. For example, a clear-cut caatinga forest site on Spodosols had 3.8 sprouts present per m^2 at 3 years and 73 per cent of the 1290 g/m^2 above-ground biomass was composed of sprouts (Uhl *et al.*, 1981a). Leaf area index was 4.8 at this time. Almost all sprouts originated from stems < 5 cm dbh. Apparently, as stems grow larger they lose their ability to sprout. Clear-cut forest sites on Ultisols at San Carlos are also dominated by sprouts and have rates of biomass accumulation close to two times greater than Spodosol sites (Uhl, unpublished data).

The importance of stump and root sprouting in the recovery of cut neotropical forests has frequently been mentioned (Ducke and Black, 1953; Emerson, 1953; Sarukhan, 1964; Snedaker, 1970; Uhl *et al.*, 1981a). Of particular importance are those species which produce root sprouts following cutting disturbances. Such species can form near monocultures for several 100 m^2 around the cut parent tree. The proclivity to sprout simplifies the succession after cutting: tree sprouts dominate the early months and simply grow larger. Tree species which are able to sprout are disproportionately represented in cut-over forest.

The ability of many Amazonian tree species to sprout may facilitate the establishment of desirable species in clear-cut sites. Indeed, we have found that some species at San Carlos even sprout readily from stem cuttings inserted into the ground. Certainly, studies of the sprouting behaviour of important Amazonian trees are worthy of attention.

If the capacity to harvest mixed hardwood stands exists, forestry based on clear-cutting by hand and the extraction of tree boles in such a way that the topsoil and associated organic matter is not excessively damaged (e.g. cable-logging) may be feasible in Amazonia, particularly if regeneration periods of a hundred years or more between harvests are the rule.

Clear-cut burned sites

Forest clear-cutting, as shown above, favours regeneration via sprouting. However, when burning follows cutting, regeneration by fast-growing successional woody species is favoured. For example, only 10 per cent of the total site stems present 22 months after the cutting and burning of a forest site on Ultisols at San Carlos were sprouts (Uhl *et al.*, 1981b). By reducing the importance of sprouting in regeneration, burning provides fast-growing successional trees and shrubs with an opportunity to colonize. The dominant tree species initially colonizing the burned Ultisols site was *Cecropia ficifolia* Snethlage. At 2 years this species had a density of 0.36 individuals per m^2 and comprised 78 per cent of the site biomass (Uhl and Murphy, 1981b). However, *C. ficicolia* began to senesce after the second year, and by the fourth year, density for this species had declined to 0.01 individuals per m^2. The mass mortality of *C. ficifolia* caused rates of wood accumulation (i.e. the difference in wood biomass for stems 72 m tall between that present at the end of a given year and that present at the beginning) to decline from 6.3 metric tons/ha in year two to 4.2 metric tons/ha in year three. By the fourth year, fast-growing, longer-lived second-growth species, such as *Vismia* spp. and *Palicourea* spp., were well established on the study site and wood accumulation for this year increased to 10.4 metric tons/ha. Because some trees present at 3 years died during the fourth year, actual wood production (i.e. wood annual increment) was close to 12 metric tons/ha for year four. Total live wood accumulation on this site at 4 years was 21.0 metric tons/ha.

This pattern of wood accumulation rate with a small peak after 2 years and a dip as *Cecropia* senesces, followed by a rapid increase as longer-lived successional species establish, appears to be typical for other burned sites in the San Carlos region. A similar dip in wood accumulation rate probably occurs after about 10–15 years when the intermediate-stage successional trees begin to senesce.

The undisturbed forest at San Carlos appears to be considerably less productive than burned-over clear-cut sites. For example, in a recently completed 5-year productivity study (Jordan and Murphy, unpublished), average wood annual increment was 6 metric tons/ha in the Ultisol forest (compared to about 15 metric tons/ha in the Ultisol burned site for the fourth year); average diameter growth for trees $\geqslant$ 10 cm dbh was 0.13 cm in the Ultisol forest (compared to an average of about 1.5 cm for trees $\geqslant$ 1 cm dbh in the burned site).

While burning has the advantage of favouring fast-growing successional vegetation, it can also result in large nutrient losses as elements in the cut slash are rendered into ash. This was shown in a San Carlos study where nutrient concentrations in drainage water were initially much greater for all macronutrients in a cut-and-burned Ultisol study site than in an adjacent forest control site (Jordan, 1981a). By 3 years the leachate water nutrient concentrations in the burned site had returned to levels similar to a control forest plot. (See extract from *Ambio* paper; pp. 152–155.)

This examination of regeneration on burned-over sites suggests that, although successional trees are more productive than forest tree species, any attempt to develop sustained-yield forestry around natural regeneration of successional species in Amazonia will have to:

(1) rely on fire to kill tree sprouts, thereby allowing fast-growing successional species to dominate;
(2) carefully consider the life histories of the successional tree species present to determine the most economical time to harvest a site; and
(3) eventually entail nutrient supplements.

Clear-cutting and tree harvest followed by plantation establishment

Tree plantation forestry is a much more intensive use of the land than either the selective harvest or clear-cut natural regeneration approaches already discussed. Although we have not studied any aspects of plantation forestry at San Carlos, this form of forest management appears to have limited promise in Amazonia. An example of plantation forestry in Amazonia is the Jari forest estates in Brazil (Briscoe, 1978). Approximately 100,000 ha of this 1.5 million ha holding have been converted to tree crops. We will use the Jari enterprise to illustrate some of the problems with tree plantation forestry in Amazonia.

The Jari project has not been a success to date because of ecological, economic, and social problems (Kinkead, 1981). The principal ecological problems have been slow growth rates, nutrient loss, and insect outbreaks. Production of above-ground dry-weight wood biomass in an experimental plantation of *Pinus caribaea* at Jari which received frequent weedings averaged 9.8 metric tons/ha each year for the first 9 years of the first pine rotation (Jordan and Uhl, unpublished data). In 13 other tropical countries, annual production averaged 18.4 metric tons/ha (Jordan, 1981b). *Gmelina*'s average yields on the best of Jari's soils are 40 per cent below target and so far off on the extremely infertile sandy soils covering 75 per cent of the plantation that these stands are being eliminated as fast as they can be cut (Kinkead, 1981). None of Jari's lands have completed more than one rotation. With each tree harvest more and more nutrients will be mined from the land. Using leaching data from cut-and-burned forest in San Carlos and nutrient content of boles and of soils at Jari, we calculated that all the potassium in the system would be depleted after two harvests. Finally, the lack of strong seasonality at the Jari site and the presence of monoculture tree crops has meant that insect and fungal outbreaks are a constant threat (Fearnside and Rankin, 1980).

Two economic aspects of forest conversion to tree plantations at Jari also deserve mention. First, Jari makes little use of the wood in the native forest when this forest is cleared to establish plantation plots. Yet, this wood represents a valuable resource. Second, given the remoteness of the Amazon Basin (and the Jari plantation), it remains to be seen if plantations there will be economically competitive with plantations growing nearer to potential markets.

Sociological impacts also must be considered in an undertaking the size of Jari.

Thousands of workers have been brought in, and their livelihood depends upon the continued growth of exotic species in a nutrient-impoverished area. If the project fails, these people may become a disruptive force to the surrounding Amazon forest.

All of this is not to say that plantation forestry is an inappropriate land-use for other areas of the tropics. Indeed, pine plantations in Central America and the Caribbean islands, where nutrients are more plentiful and where logistic facilities are reasonably well developed, have been shown to be a good land-use alternative (Whitmore and Liegel, 1980).

CONCLUSION

Throughout this chapter we have emphasized the small stocks of nutrients in the Amazon forest ecosystem. This paucity of nutrients is the fundamental constraint to long-term production forestry in Amazonia. Forests can probably be managed for long-term sustained yield in Amazonia if great care is taken, but costs must necessarily be high because the land will have to be cleared and managed carefully, and fertilizer supplements will probably be required at intervals. Economic analyses may, indeed, reveal that it makes better economic sense at present to pursue forestry alternatives in more accessible, less nutrient-impoverished areas.

THE EFFECT OF HUMAN ACTIVITIES

R. Herrera, C. F. Jordan, E. Medina and H. Klinge

(Extract from a paper in *Ambio*, 1981, and reprinted by permission of Ambio, Royal Swedish Academy of Sciences)

As one of the experimental treatments in our projects we selected plots on forest sites similar to that where the nutrient cycle was studied. These plots were felled in August, 1976; after drying they were burned in December of the same year. One plot was planted with cassava and pineapple as the main crops with a few plantains and cashews. The second plot was not cropped, and secondary successional vegetation was allowed to become established after the burn. Nutrient concentrations in the two experimental areas and in a control were monitored by sampling water from lysimeters installed below the root zone (Jordan, 1968). Water samples collected weekly were analyzed for nutrient elements. Figure 2 shows the comparison of nutrient concentra-

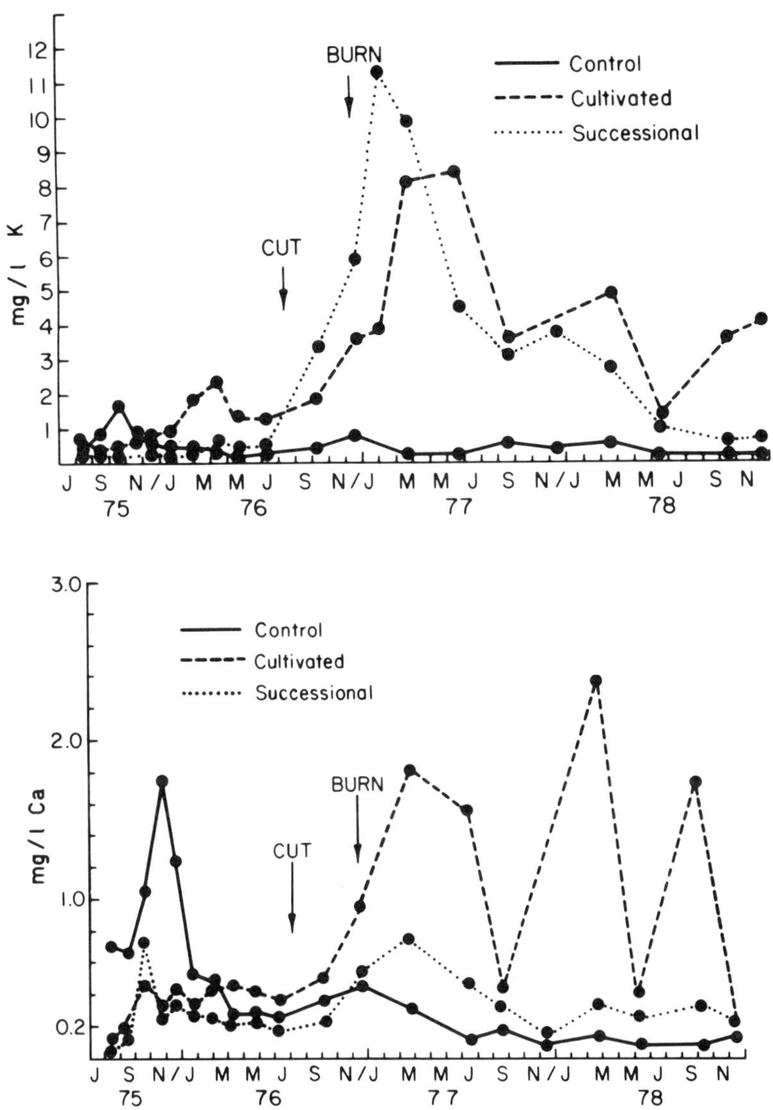

Figure 2 (Top) Concentration of potassium in soil water percolating to a depth of 12 cm in the control, cultivated, and successional areas. Points represent averages of 12 lysimeters in control area and 6 in each experimental area. (Bottom) The same for calcium. Reproduced by permission of *Ambio*, Royal Swedish Academy of Sciences

tions between the control and the two experimental treatments. For potassium, an increase in concentration was observed for both the successional and the cultivated plots, starting right after cutting and peaking between one and six months after burning. No significant difference could be observed between treatments. Calcium concentrations show larger variability in the control; after the plots were burned a sharp increase in concentration was observed for the cultivated plot. The successional plot also showed an increase in concentration, but it was much less pronounced than for potassium. Magnesium and nitrate–nitrogen concentrations increased in a manner similar to potassium, while ammonium–nitrogen behaved like calcium, with greater increases in the cultivated plot (Jordan, in press). The rate of recovery of the forest after disturbance depends on how much of the nutrients are irretrievably lost for the successional vegetation. In our experiments calcium and ammonium–nitrogen seem to be lost more rapidly under cultivation than under successional vegetation. The development of the dense root network needed to capture these and possibly other elements before they are lost to runoff is considered an important factor controlling the rate of recovery.

Other factors affecting recovery are the length of time a plot is cultivated and the length of the fallow period before cultivation is repeated. These seem to be critical in areas where—either because of cultural causes or simply because a larger population is being fed with the same land—the farmer crops the plots for longer periods or permits shorter fallows. In the vicinity of San Carlos, we believe that the maintenance of artificially large populations with restricted mobility increases the pressure on land-use beyond the traditional use by native populations elsewhere. To tackle this problem we have recently started a research program on human ecology with emphasis on cultural adaptations and resource use by both traditional groups and acculturated town people.

Simultaneously we have been monitoring soil fertility in shifting cultivation plots around San Carlos. A sequence of plots ranging in age from one to eight years under cultivation has been selected. Additionally a very old plot which we estimate had been cropped repeatedly for at least eight years and then abandoned for two, was also identified. Figure 3 shows the levels of total nitrogen and phosphorus and exchangeable calcium and potassium in the soil samples of this series of plots arranged according to their approximate age. An initial increase in calcium and phosphorus was observed, while nitrogen and potassium fell below the original values under forest. This difference can be ascribed to the solubility of nitrate–nitrogen and potassium in the soil after the disturbance. The slow decomposition of the remaining biomass of the forest, particularly stumps and roots, could account for the increase of calcium and phosphorus. The traditional practice of abandoning the land after three years or less seems to be based on the existence of temporarily high nutrient levels either in the soil water or in the exchangeable complex in the soil. After this, one or more nutrients necessary for the crops might become limiting and in consequence yields drop. In addition, though this is seldom, the recuperation of fertility of a site through fallow might proceed very slowly if the land is allowed to reach too low nutrient levels, although this may not be a consideration with the farmers.

CONCLUSIONS

The functioning of tropical rain forests with limited soil nutrients depends largely on input by rain and the ability of the forest plants to use this source of nutrition. The forest shows very well developed mechanisms for efficient re-cycling of the nutrients stored in its biomass so that losses are minimal. These mechanisms are of high adaptive value and can be identified both at the structural and physiological levels. The forest

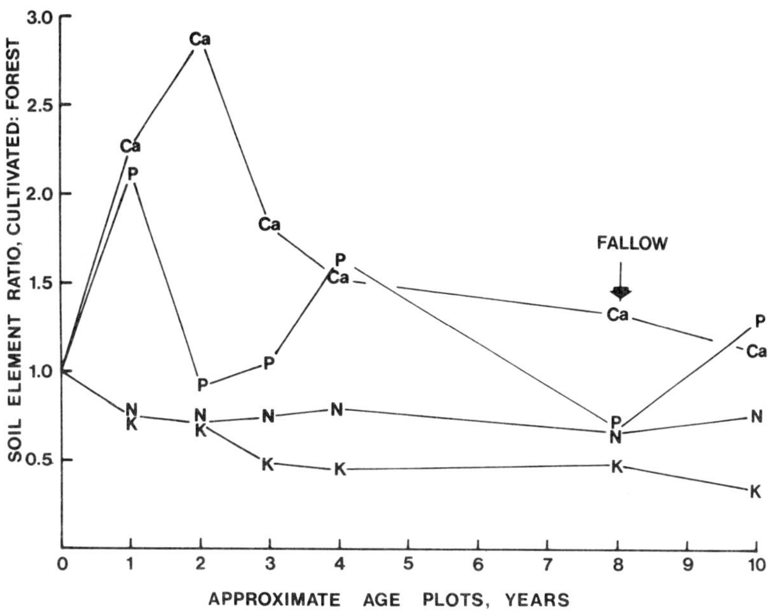

Figure 3 Course of soil fertility in shifting cultivation plots of different ages under cultivation. Nitrogen and phosphorus are total organic, and calcium and potassium are exchangeable. Values given are ratios between cultivated soil and soil under forest for each nutrient (identified in respective curve by chemical symbol). Reproduced by permission of *Ambio*, Royal Swedish Academy of Sciences

trees are able to exist on reduced nutrient budgets compared with forest growing on more fertile locations. It is considered more pertinent to compare forests growing under 'oligotrophy' versus those growing under 'eutrophy' than the more usual contrast of tropical versus temperate ecosystems (Jordan and Herrera, in press).

The nutrient-conserving mechanisms are very fragile and when the forest is disturbed they cease to function. The soil, with its low retention capacity, is impoverished after a brief period of fertilization due to the sudden release of nutrient in the biomass. This brief period of fertility is used by shifting cultivators to produce crops. But even this system, which was developed by the native populations in close contact with the environment, can induce severe losses if forced to fit a market-oriented economy or to feed larger populations. Serious efforts should be directed to assess the impact of large scale disturbances of forest land in Amazonia and to investigate alternative management practices that could assure sustained yield agriculture with appropriate technologies (Janzen, 1973; Greenland and Lal, 1975).

REFERENCES

Beek, K. J., and Bramao, D. L. (1968). Nature and geography of South American soils. In *Biography and Ecology in South America* (Eds E. J. Fittkau, J. Illies, H. Klinge, G. H. Schwabe and H. Sioli), vol. 1, pp. 82–112. Junk, The Hague.

Briscoe, C. B. (1978). Silviculture in plantation development. In *Proceedings of Conference on Improved Utilization of Tropical Forests*, pp. 125–139. Madison, Wisconsin. Forest Products Laboratory, Forest Service, USDA.

Burgess, P. F. (1973). The impact of commercial forestry on the hill forests of the Malay Peninsula. *Bio Indonesia*, **1**, 17–23.

Collardet, J. (1976). Processing hard-to-process and lesser used species. *Unasylva*, **28**, 93–105.

Ducke, A., and Black, G. A. (1953). Phytogeographical notes on the Brazilian Amazon. *An. Acad. Bras. Cienc.*, **25**, 1–46.

Emerson, R. A. (1953). A preliminary survey of the milpa system of maize culture as practised by the Maya Indians of the northern part of the Yucatan Peninsula. *Annals of the Missouri Botanical Garden*, **40**, 51–62.

Erfurth, T. (1976). Toward the wider use of tropical wood products. *Unasylva*, **28**, 119–26.

FAO–UNESCO (1971). *Soil Map of the World*, vol. IV, *South America*. FAO, Paris.

Fearnside, P. M., and Rankin, J. (1980). Jari and development in the Amazon Basin. *Interciencia*, **5**, 146–56.

Fittkau, E. J., Junk, W., Klinge, H., and Sioli, H. (1975). Substrate and vegetation in the Amazon region. In *Vegetation and Substrat* (ed. H. Dierschke), pp. 70–93. Vaduz, J. Gramer.

Furch, K. (1976). Haupt- und Spurenmentallgehalte Zentral Amazonischer Gewassertypen. *Biogeographica*, **7**, 27–43.

Furch, K., and Klinge, H. (1978). Towards a regional characterization of the biogeochemistry of alkali- and alkali-earth metals in northern South America. *Acta Cientifica Venezolana*, **29**, 434–44.

Grainger, A. (1980). The state of the world's tropical forests. *Ecologist*, **10**, 6–54.

Greenland, D. J. and Lal, R. (1975). *Soil Conservation and Management in the Humid Tropics.* J. Wiley and Sons, Chichester.

Herrera, R. (1977). Soil and terrain conditions in the International Amazon Project at San Carlos de Río Negro, Venezuela: correlation with vegetation types. In *Transactions of the International MAB–IUFRO workshop on tropical rain forest eco-systems research*, pp. G, 182–7. Hamburg–Reinbek.

Herrera, R. (1979). Nutrient Distribution and Cycling in an Amazon Caatinga Forest on Spodosols in Southern Venezuela. PhD thesis, University of Reading, Reading, England.

Herrera, R., Jordan, C., Klinge, H., and Medina, E. (1978a). Amazon ecosystems: their structure and functioning with particular emphasis on nutrients. *Interciencia*, **3**, 223–31.

Herrera, R., Merida, T., Stark, N., and Jordan, C. (1978b). Direct phosphorus transfer from leaf litter to roots. *Naturwissenschaften*, **65**, 208–9.

Herrera, R., Jordan, C., Medina, E., and Klinge, H. (1981). How human activities disturb the nutrient cycles of a tropical rain forest in Amazonia. *Ambio*, **10**, 109–14.

Janzen, D. (1973). Tropical Agroecosystems. *Science*, **182**, 1212.

Jordan, C. F. (1968). *Soil Science*, **105**, 81.

Jordan, C. F. (1981a). Nutrient leaching from agro-ecosystems in the Amazon Basin, and implications for the recovery of the forest. *Proceedings of the Fifth International Symposium in Tropical Ecology*, Kuala Lumpur, Malaysia.

Jordan, C. F. (1981b). Productivity of tropical rain forest ecosystems and the implications for their use as future wood and energy supplies. In *Ecosystems of the World*, vol. II: *Tropical Rain Forests* (ed. F. Golley). Elsevier, Amsterdam (in press).

Jordan, C. F., and Herrera, R. (1981). 'Tropical rain forests: Are nutrients really critical?' *American Naturalist*, **117,** 167–80.

Kartawinata, K. (1975). The exploitation of natural forests and natural areas development. *Bio Indonesia*, **1,** 17–23.

Kinkead, G. (1981). Trouble in D. K. Ludwig's jungle. *Fortune*, 20 April, 1981, pp. 102–17.

Klinge, H., and Herrera, R. (1978). Composite root mass in tropaquods under Amazon caatinga stands in southern Venezuela. *Proceedings of the Fourth International Symposium of Tropical Ecology*, Panama, vol. 1, pp. 395–420.

Kyrklund, B. (1976). Paper from mixed tropical forests. *Unasylva*, **28,** 86–92.

Lawton, R. M. (1978). The management and regeneration of some Nigerian high forest ecosystems. In *Tropical Forest Ecosystems*, pp. 580–88. UNESCO/UNEP/FAO, Unipub, New York.

National Academy of Sciences (1980). *Conversion of Moist Tropical Forest*, National Academy of Sciences, Washington, DC.

Palmer, J. R. (1977). Forestry in Brazil–Amazonia. *Commonwealth Forestry Review*, **56,** 115–30.

Prance, G. T. (1978). The origin and evolution of the Amazon flora. *Interciencia*, **3,** 207–22.

Richards, P. W. (1952). *The Tropical Rain Forest.* Cambridge University Press, Cambridge.

Sanchez, P. (1976). *Properties and Management of Soils in the Tropics.* John Wiley & Sons, New York.

Sarukhan, J. (1964). 'Estudio sucesional de una area talada de Tuxtepec, Oaxaca'. Publicacion especial, *Instituto National de Investigaciones de Forestales de Mexico*, **3,** 107–72.

Snedaker, S. (1970). *Ecological Studies on Tropical Moist Forest Succession in Eastern Lowland Guatemala.* PhD thesis, University of Florida, Gainesville, Florida.

Stager Carvajal, P. F. (1978). Little known woods of the Brazilian Amazon. In *Proceedings of Conference on Improved Utilization of Tropical Forests*, pp. 163–82. Madison, Wisconsin. Forest Products Laboratory, Forest Service, USDA.

Stark, N., and Jordan, C. (1978). Nutrient retention in the root mat of an Amazonian rain forest. *Ecology*, **59,** 434–7.

Uhl, C., and Murphy, P. (1981a). Composition, structure, and regeneration of a tierra firme forest in the Amazon Basin of Venezuela. *Tropical Ecology* 22:219–37.

Uhl, C., and Murphy, P. (1981b). A comparison of productivities and energy values between slash and burn agriculture and secondary succession in the upper Río Negro region of the Amazon Basin. *Agro-Ecosystems*, **7,** 63–83.

Uhl, C., Jordan, C., Clark, K., Clark, H., and Herrera, R. (1981a). Ecosystem recovery in Amazon caatinga forest after cutting, cutting and burning, and bulldozer clearing treatments. *Oikos*, (in press).

Uhl, C., Clark, K., Clark, H., and Murphy, P. (1981b). Early plant succession after cutting and burning in the upper Río Negro region of the Amazon Basin. *Journal of Ecology*, **69,** 631–49.

Wambeke, A., van (1978). Properties and potentials of soils in the Amazon Basin. *Interciencia*, **3,** 233–42.

Went, F. W., and Stark, N. (1968). 'Mycorrhiza'. *Bioscience*, **18,** 1035–9.

Whitmore, J. L., and Liegel, L. H. (1980). Spacing traits of *Pinus caribaea* var. *hondurénsis. USDA Forest Service Research Paper* SO–162. Institute of Tropical Forestry, Río Piedras, Puerto Rico.

Socio-economic Effects and Constraints in Tropical Forest Management
Edited by E. G. Hallsworth

15

The human face in forest management

L. J. WEBB
School of Australian Environmental Studies, Griffith University, Brisbane, Australia

INTRODUCTION

Committees have been defined as 'organisms that excrete paper'. Every cellulose explosion surrounding a conference consumes more hectares of forest. I need not remind you of the enormous amount of literature that touches on our problems, and that pours daily from august international and national institutions, and specialists and laymen of all kinds. This mass of words eventually dulls our senses. At the 8th World Forestry Congress at Jakarta in October 1978, at the International Society for Tropical Ecology meeting at Kuala Lumpur in May 1979, and at various Man and the Biosphere (MAB) committee meetings in Australia and abroad—I hear the same refrain; it has all been said before, with varying degrees of eloquence, demagogy, honesty, impertinence, or downright inaccuracy.

Yet it is a significant fact, and highly relevant at the time of writing, that of the many metres of bookshelves occupied by such literature, perhaps less than a metre of books attempts to integrate the physical, biological, and socio-economic aspects of our problem. Why this is so is in itself a tempting (and dangerous) subject for scientific enquiry.

We all tend to take so much for granted. When we say 'forest management', our first thoughts lean towards the operations of a forester trained in some variation of the Western tradition—operations designed to extract wood that is mostly destined for the developed countries. Yet forest management has been, and still is in many areas, undertaken by local people without any academic training; people who perceive and utilize the environment differently; people who have managed their forests for millennia, and who have complex traditions that even now attract little scientific attention other than of an external and exploitative kind.

Nevertheless professional foresters have become aware of these dilemmas, and some are now urging a 'New Forestry Policy' (Qureshi *et al.*, 1980). This is a theme well worth exploring, so that the goals of forest management can be clarified by reference to the needs of the local people, whether they be complementary to, or in conflict with, the needs of other people who want to use the forests.

I have consequently decided to restrict myself to a discussion of one theme. More detailed data about several Case Studies in Papua New Guinea and Northern Australia are available in the literature (Webb, 1977; De'Ath, 1978a, 1980; Anon, 1976, 1977; Lamb, 1977). I offer these notes to indicate some of our current and developing ideas. If we can collectively establish some new approaches, however tentatively, we may then plan effective research and training to resolve some of the very difficult problems dealing with the human face of forestry in the tropics and subtropics, although these problems are by no means restricted to the developing countries.

THE PLACE OF PEOPLE IN ECOLOGICAL RESEARCH

From the following examples, I suggest that a new kind of ecological research is emerging that emphasizes human values as well as describing biophysical factors. It requires specific recognition, definition and promotion.

Case A: The Madang timber project

In the Gogol Basin, near Madang, Papua New Guinea, a large woodchipping operation has been undertaken in a lowland rainforest by a Japanese company, Japan and New Guinea Timbers (JANT) signed in Canberra before Papua New Guinea became independent, but with the approval of Papua New Guinea people who were to form the first government of the independent state.

It began with logging in 1969. Clear-felling for woodchips began in 1973, the chips being exported to JANT's parent company, Honshu Paper. The industry was the first of its kind in Papua New Guinea, and the first in the South-east Asian region to use rainforest tropical hardwood chips to make pulp for paper. The study of the impact on the local people, who were at that time subsistence agriculturalists, is consequently of great interest, regionally and worldwide.

This new use of mixed hardwoods from tropical rainforests to produce paper pulp meant not 'selective logging', as previously (although that was often an indiscriminate operation), but clear-felling of practically every tree, except for some of unsuitable shape for de-barking, or that had been marked by the local people as food trees (e.g. supply of nuts). Thus the operation represented an entirely new and revolutionary change in the lifestyle of the local people, who had previously known only a few hectares at a time, in a mosaic pattern, to be clear-felled for subsistence agriculture. Now they were to witness and live among the results of clear-felling operations moving 'on a face' through the tall luxuriant rainforests of the alluvial terraces, and reaching up the foothills into the rather poorer forest types which were more patchily logged. The results have not been quite as had been hoped, and emphasize the need to consider people as part of the 'ecology'.

From this work in the Gogol Basin, the social anthropologist Colin De'Ath suggests the term 'action research'. We may indeed wish to call it 'action–reaction–interaction research', for that is what is meant. It embraces 'bottom-to-top' as well as 'top-to-bottom' flow of information and implementation. Bottom-to-top, the active consultation with and participation of the local people affected by any development proposal and operation, is regarded as the primary goal of such research, and must involve training and education on the people's own terms.

For example, in the Gogol, De'Ath works from the assumptions that:

(1) problems are imported by large outside institutions;
(2) the people have a viable relationship with nature;
(3) as outsiders we can learn from this relationship especially about the weaknesses of our own equations with nature;
(4) it is incumbent on the researcher to obtain and transmit crucial information about the encroaching institutions and their purposes to the Gogol people;
(5) if the government behaves in the same way as large resource-destroying corporations it should not be free from criticism;
(6) if the researcher and the people see the need for buffer or defensive institutions or strategies they should not be averse to implementing them;
(7) at some stage the researcher will become useless in terms of being able to continue to gather information about large institutions for the villagers unless he becomes, probably inappropriately, a political figure in the larger political system;
(8) in the whole process the information gathered by the researcher should be given to the people first—information *is* power;
(9) the whole process should be reflexive and what happens in the research—action phase should be carefully recorded to validate what was probably a pattern in prior colonial relationships.

These ideas were explained further (Samol and Webb, 1978) at the MAB Workshop, October 1978, at Jakarta, and then developed by De'Ath as elements of ideal 'social impact research' in his recently published monograph (De'Ath, 1980). In an earlier draft, not included in the publication, he spoke frankly of what he called 'The biases of the author'. He noted the complexity of the woodshipping project at the Gogol and its implications, and expanded his earlier ideas about 'action research'. His notes are published here as a personal communication.

The project has been criticized from a number of positions: from the point of view of the throwaway nature of the end-product—packaging; from the perspective of the ecologists and environmentalists; from the economics or costs and benefits point of view; from the aspect of the nationalists who want to see both secondary and extractive industries established in Papua New Guinea; from the scientific point of view (does the research being done under the auspices of a 'Man and Biosphere' programme meet the needs of those with a direct interest in the project?); and from the point of the people in the Trans-Gogol who have had direct experience in the project's impact.

Within the study a number of assumptions have been made. These assumptions should be made overt at the beginning of the study so that readers are under no illusions as to the biases of the author.

1. The local people in such a large-scale project as this should not be regarded as a problem. The people who create problems are the decision-makers in Port Moresby, Canberra, Madang and Tokyo. Subsistence agriculturalists are among some of the least offensive or problem-provoking people left in the world.
2. This project should not be looked at in isolation in terms of either time or space. For the purpose of this exercise, the project has been examined in the context of a colonial history. The motives of the Jant personnel have not been greatly different from those of the early missionaries, government and other personnel. In this kind of context the Jant operation is seen as perhaps accelerating a process which was already occurring. Thus historical continuities are emphasized in the relationships between Trans-Gogol villagers and outsiders.
3. It has been presumed that to maintain elites and urban populations in a country such as Papua New Guinea surpluses must be generated in hinterland areas from agriculture, from large rural extractive industries or from overseas aid. The relationship of the Trans-Gogol to Madang, for example, is examined from this point of view.
4. In anthropological literature and other social and ecological science literature there is still a tremendous ambivalence about the quality of life in subsistence, self-contained communities and whether such communities were/are dynamic or static in terms of growth and change. I do not wish to become involved in this argument in such a short work. My assumptions have come to be:

(a) Life was not as nasty, mean, brutish or short as the missionaries, for example, would have us believe. However, because of sickness and natural disasters it was far from ideal. In the inland areas of the Gogol Timber Rights Purchase area (TRP), the present quality of life may be inferior to what it was prior to western contact.
(b) Life in the village was not programmed as exquisitely as the functionalists would have us believe. If functionalism had worked well over time, the population in the Trans-Gogol might not have been as sparse as it is. Better or more efficient use might have been made of plentiful local resources, despite the effects of natural disasters on the carrying capacity of the land.
(c) The people have not been, or are not now, total opportunists or pragmatists. They have tended to be satisfied with reasonable access to local and traded resources and do not welcome change for its own sake. Additionally, a long history of contact with the colonialists and cultists has made them very wary of accepting the imperatives of outsiders.

(d) Because of the historical feedback implicit in (c) and because of their historical conservatism, they have made some fairly healthy judgements about the costs and benefits of current involvements with outsiders. For example, they have rejected local government, missionaries, cultists, national government and now Jant because of their correct assessment that these institutions are interested only in asymmetrical or exploitative relationships.

(e) Although their current assessment of their situation *vis-a-vis* outsiders is correct in terms of exploitation, their methods of coping with outside-initiated change may not be. Additionally, their information about outsiders and their motives is inadequate. Thus strategies of reversion to reconstructed ancestral history, including cultism, isolation in fragmented clan areas, migration to towns (this strategy is problematical because in Papua New Guinea resources *are* concentrated there), and less schooling (also problematical) are inadequate to cope with change. Outsiders know how to establish and operate large predatory institutions and it would seem as though villagers must also create such institutions to confront outsiders and to survive. However, whether they do or do not do this, the ultimate cost will be high: possibly cultural extinction if they do, or survival in a low quality physical environment if they do not.

(f) In such a situation as that outlined in (e) the villagers may ultimately become schizoid—damned by both strategies of imitation and withdrawal. At present these kinds of considerations make the Trans-Gogol people apprehensive about the future and nostalgic about the past with its powerful deities. In such an uncertain situation, the current quality of life, certainly on a cognitive level, is probably quite low.

5. Although this research was funded by the Office of Environment and Conservation of Papua New Guinea, I have taken the responsibility of trying most of the time to see the changing world from the Trans-Gogol people's own perspective and have tried also to supply information which I hope will be helpful to them. At this late stage in the project the study reads more like an obituary than a prescription for citizen action. However, perhaps other affected groups will learn something from it so that they can at least ask the right questions prior to the implementation of a large-scale rural development project. To me it has become painfully obvious that the people should at the outset of any national project be the recipients of enough funds to hire their own advocates and experts, particularly maverick economists.

In the future, the existence of provincial government in all provinces will very much alter the kinds of tactics developers and affected groups must use. Whether this extra layer of bureaucracy will bring government closer to the rural people or become yet another greedy or cuckoo-like institution hastily scrambling for royalty titbits, remains to be seen. From the experiences of other countries near to Papua New Guinea, it would seem that local and regional governments are even more insensitive in the disposition of resources than are national governments. It is my own wish that this will not happen. However, urban Papua New Guinea has become dependent on outsiders for manufactured imports and it would seem, regardless of how government is structured, that what has happened in the Trans-Gogol will happen to the nation as a whole. Let us hope that not all forests will be made into throwaway commodities nor future costs to local people be treated as project externalities.

The time perspective and the methodology

The trajectory of social change commences at a hypothetical point some 50,000 years B.P. At that time the newly arrived people in Papua New Guinea were hunters and gatherers and belonged to small political units which were self-contained. They were in a relatively homeostatic state in terms of their natural environment. As one moves up the change trajectory to the present it can be observed that western influence is a very recent occurrence. At present, 1978, the people are no longer ecosphere people but rather biosphere people. They have become dependent on, and contributors to, a global system of commodity and resource exchanges. This study documents the more salient features of changes since 1871 with a heavy emphasis on the effects of JANT on the change trajectory.

The study, which has occupied about one year with only about six months spent in the field in the Trans-Gogol was too short to capture great cultural depth. However, its purpose was to *identify the etiology of current problems and not just the effects*. It was, therefore, considered important to study the people and institutions which imported problems into the Trans-Gogol.

The model and methods used for the study have several dimensions. An attempt has been made to elicit historical continuities, particularly in terms of colonial relationships. An assumption was made at the outset of the study that most large-scale institutions introduced into Papua New Guinea relate to villagers in much the same way. Their specific methods, strategies and ideologies might be different but there is, or has been, a great deal of congruence in their goals and ultimate effects. The material for the description of the historical development of institutions was gathered from institutional records and oral histories. The people's perceptions of their own history, myths and of their contact with such institutions were also gathered orally. The study is synchronic but attempts have been made to have a time depth by using the known dates of historical events on which to hang unrecorded happenings and events. In brief, the following techniques were used to elicit information:

(1) library, archive and current government file searches;
(2) participant-observation in JANT operations, in the Trans-Gogol villages and in church and government activities;
(3) questionnaires and letters to knowledgeable persons.

At the conclusion of the study, and when the write-up had been completed, drafts for comment and correction were circulated to a number of interested organizations including JANT and the Office of Forests. The study was also summarized in Pidgin English for radio in the hope that Trans-Gogol national audiences would benefit from the information gathered.

Bitterness, frustration and even anger may be inferred from the study. This is because the study reflects the mood of the Trans-Gogol people who are literally on the cutting edge of change. It is expected that a good deal of corporate effort will go into denigration of the study through questioning the accuracy of detail and bias rather than into redressing the broader problems of inequality, unequal international dependency relationships, rural–urban differences and village participation in decisions on matters affecting themselves.

The need for this study might have been obviated if a careful record of the social impact had been kept of, say, the timber and gold extractive industries in the Wau-Bulolo valley. Even to the casual outsider it is obvious that the people from these two areas have not benefited materially from resource exploitation over a period stretching back nearly fifty years.

The impact of other kinds of research and assessment

A 'Man and Biosphere' (MAB) programme was officially launched by UNESCO in 1970. Essentially this new programme was an outgrowth of, or successor to, the International Biological Programme. This latter programme impinged on the then

Madang District (now Province) with equivocal results. The MAB research in Papua New Guinea focusses on the Trans-Gogol and has been operative since 1974. The Trans-Gogol people were never asked whether they wished their area to be a focus for research nor whether they wished themselves to be experimental or research subjects. This oversight might be attributed to the fact that the research programme was planned and co-ordinated by physical scientists. Even the ecologist, amongst the twenty-five august researchers wishing to do research in the Trans-Gogol, overlooked one of the most important components of interdisciplinary research: human co-operation. Nor were the people made aware of what the impact of such research would be. For example, they still do not know that about K160,000 per year is being spent in their area on research. One of the reasons for their not seeing the magnitude of the research mountain being built is that researchers, unlike guano-producing birds and bats, do not leave great piles of valuable droppings near where they eat. On the contrary, their role, like that of JANT, is extractive. Their mountain is being built elsewhere. Their mission has been to collect blood, faeces, animals, soil, flora, skin and information. If the people had perhaps understood that the hasty visits by researchers and research fund administrators each year cost approximately eight times as much as they got for their royalties from 1974 to 1976, their co-operation with peripatetic scholars might have been less readily given.

It is significant that at the international MAB workshop held in Madang in March 1978 the people complained about the lack of feedback from research and about the commuter styles of most researchers based in Madang and elsewhere. The foresters, perhaps huffily, felt that this criticism did not apply to them, because they had built a facility in the Trans-Gogol and because Forest Officers have a 'day-to-day presence' there.

Given the indifference of some researchers to the welfare and interests of the Trans-Gogol people, it would not be surprising if the people later either denied them entry in the future or placed a bounty on their research.

From the stance of the typical scientist some Trans-Gogol research is faulty too for the following reasons:

(1) it is after-the-event research without enough being known about conditions *prior* to the intervention by JANT;
(2) the time span of most research projects is too short to fit the aspirations of the researchers, or conversely, the aspirations of the researchers make their research too transitory;
(3) reductionistic research tends to take priority over more comprehensive, integrated research;
(4) there is more emphasis on taxonomic, rather than dynamic, systems-oriented research;
(5) the relationship between so-called applied and pure research has not been thought through;
(6) research questions have been generated only by westerners, professional scientists and large-scale institutions—local people have been excluded from this interest-eliciting process;
(7) the programme has not been free from academic carpet baggers—those who have research methods or techniques to hawk and who have forced problems to fit their methodologies.

For a detailed evaluation of research in the Trans-Gogol the reader is referred to another work of the author. (De'Ath, 1978a, 1978b).

These are indeed bitter and chastening observations. For De'Ath (and I must include myself) research can never be value-free. It is rare to encounter such forthright ideas before they have been 'to the laundry', and I hope that we will benefit from them.

George Seddon (1978) took a somewhat milder view, but did not conceal the problems:

The Gogol venture is of interest for several reasons:
(1) It is a case-history of a development project in a poor country (one in which the host has had to pay much higher costs than were originally reckoned with), and of a project in a little understood lowland tropical rainforest.
(2) It is an exercise, although a difficult one, in total evaluation; i.e. what is the sum of the costs and benefits of such an operation for all concerned?
(3) Further, it surely has a profound introspective interest for Australians as part of our colonial legacy to Papua New Guinea. A nation shows its basic decencies and shortcomings with embarrassing clarity in its colonial record, and there is much to be learned about Australian society in studying the Gogol.
(4) Finally, there is the interest of preventive therapy; by identifying the mistakes of the Gogol we may hope not to repeat them; and it might even be possible to mitigate some current ill-effects on the people who still live there among the stumps and the secondary growth . . .

Finally, there are the intangibles, and this brings us to the people [*sic*] of the Gogol.

The role of the villagers: While the tale of the royalty payments is a sorry one—in some villages, as little as only 2 Kina per head, the equivalent of 6 bottles of Coca-Cola—it is also one of learning from experience, and the present system of consolidating part of the payment into a trust fund which can be used as a source of capital for communal commercial enterprises is a good one. These lessons of the Gogol have been learned.

The benefits to the people of the Gogol have been the roads and the provision of local employment, which has brought some cash into the villages. Everyone now has access to a transistor radio, for example, and this has improved communications greatly, as Radio Madang is used freely for messages of many kinds. The struggles of the Gogol people have also heightened their political consciousness, and begun to create a regional identity.

Finally, the Madang Project may be seen to increase the range of *choice* available to the villagers. In theory, no one *has* to work for JANT, and each man therefore appears to have a choice between joining the cash economy, or sticking with subsistence agriculture in the village, if that is what he wants. In practice, this is not an unconstrained choice—all the pressures are one way.

What have the villagers lost? Above all, their forest, and what it has meant as a source of food, building material and personal adornment.

The debate that is central to the Madang Timber Project can have no simple resolution; the questions raised by the Gogol are the same dilemmas that beset thinking and self-questioning Western societies, but sharpened because the alternative paths may seem equally available in Papua New Guinea.

It is part of the debate between what have been called 'ecosphere people' who applaud small, largely self-contained systems that conserve environmental resources, as against the global technologists, who think in terms of large interlocking systems—often for genuinely humanitarian reasons, knowing well that an ethnic prison, which used to be called a ghetto, is not always comfortable, despite its internal coherence and sharp cultural identity.

At least one major lesson of the Gogol is very clear, and that is that environmental analyses are incomplete—almost pointless—unless there is an understanding of the people who live there, of their needs, aspirations, perceptions of their environment, of themselves, their past, present and future, of their economic imperatives and their political systems at every level from the local and provincial to the national—of all the forces that modulate choice and change.

This lesson is the one that Australians especially should take to heart, since the initial decision to log the Gogol was taken in Camberra, without due regard to its social consequences.

A second lesson is that the economic analyses of such projects should be comprehensive. Many of the costs of this enterprise are still emerging, and the sketch of costs and benefits given in this paper is no more than an attempt to show the proper range of such an enquiry.

If adequate ecological safeguards had been built in to the original agreement, and the real costs of reforestation and of the level of government support through its agencies had been clearly displayed at the outset it is doubtful that either JANT or the Government would have considered this a financially attractive proposition.

The third major lesson is that land-use plans consequent on logging must be more than a paper exercise: they must be realistic, in the sense of being well based on knowledge of land capability, of markets, and of the human capacity and aspirations in the region. The Gogol land-use plans meet none of these criteria, and have therefore not been realised.

Jonas (1976) earlier emphasized that 'what is fundamentally the most important aspect' of a forest ecosystem is 'the people who inhabit it and own it', and denounced the exclusion of the wishes and welfare of people 'when assessment is being made of socio-economic–ecologic strategies'. In my consultancy report on ecological considerations and safeguards for the woodchipping operation at the Gogol, I had also mentioned many of these problems, and stressed the need to integrate sociological with ecological research aimed at human welfare. My draft report had been submitted to several groups of people, and in the final report (Webb, 1977) I wrote:

The reception of the draft report varies depending on whether the readers were professional foresters or not. Those outside the forestry profession generally approved of the attempt to assemble the data into a coherent picture integrating the ecological and human aspects, and sympathized with the writer's problems in making recommendations on such a comprehensive basis. On the other hand, foresters, and especially those in the PNG Forests Department, were generally critical of the recommendations on the grounds that they were unrealistic in operational and economic terms, that they had exaggerated the 'non-wood' values of the forest environment at the expense of the 'development' values to be gained, that present ignorance of the environmental effects of total logging did not justify a 'do nothing' approach, and that the stand taken against extensive total logging in a given land system left 'no grounds for compromise' and discredited other alternatives.

Here then developed a ready-made situation which exposed the gulf (derived from differences in experience, outlook, and commitment) between the 'idealistic' ecologist concerned with the conservation of all forest values, and the 'pragmatic' forester responsible for supervising the production of wood required by industry and for economic development. If the ambitions of MAB Project No. 1. are to be realized, and if ecological principles are to have any substantive influence in planning and management it is essential from a scientific as well as a practical point of view to try to bridge the gulf. This revised report has been written in this spirit.

One conclusion which has clearly emerged is that only *practical trial* to reconcile ecological, political, and technological requirements in *specific situations* will dispel the cloudy generalizations which characterize so many modern pronouncements on environmental management and research.

Peter Driscoll, a zoologist from the University of Queensland, who recently spent 2 years in the field at the Gogol, writes (personal communication):

The past influence of the missions and government in the Gogol made it possible for the government to purchase the timber rights from the local people. The various villages were able to make a joint decision to allow their forests to be logged and people could visualise the possible benefits which would be derived from cash for timber and roads to Madang. However, amongst the locals there were some outspoken critics of the plan and their number have increased over the years. Now there is general discontent about ill effects of the logging which residents did not foresee, e.g. the decrease in birdlife (Table 1).

People's expectations of the benefits of the logging were either totally unrealistic or have not been fulfilled because the government underestimated the difficulties of further development in the area.

A major complaint by the people has been the seemingly small payment made for their timber. The method of payment has been revised several times by the government at the people's request. Total yearly payments now amount to between K50,000 and K100,000 but once it is distributed amongst 6,000 people few are completely satisfied. The construction of roads is greatly valued by locals who see them as a major

Table 1 Results of questionnaire (in Pidgin) circulated to over 150 people in 3 villages in the Gogol Valley. Listed are the percentages of replies giving affirmative answers to questions.

Eaten wild birds at least once during the preceding 2 weeks:	37%
Eaten eggs of wild birds at least once during the preceding 2 weeks:	37%
Owns decorative bird plumes:	45%
Plumes valued at between K1 and K2:	26%
Plumes valued at between K5 and K10:	60%
Plumes valued at between K30 and K100	14%
Populations of only a few birds will be decreased by the logging:	41%
Populations of many birds will be decreased by the logging:	59%
Given time for the forest to regenerate:	
Bird populations won't recover:	41%
Populations of some species will recover:	38%
Populations of all species will recover:	21%
Level of concern about the ill effects of the logging:	
Unworried:	5%
Moderately worried:	18%
Very worried:	72%

key to the future. On several occasions they have staged road blockades to stop logging operations and force the company to locate roads closer to villages for easier access. Quarrels over road construction have usually been resolved directly between the village people and the logging company under considerable tension.

The people have been learning the hard way about negotiating with the company and government to defend their position. They have been vulnerable through their ignorance of such things as the value of river gravel for roads and the disturbance to waterways caused by its removal. If you take the view that long term benefits of the project will inevitably outweigh the problems created for the people then present compensation for the timber may seem sufficient, especially as the company claims the deal is the best they can offer. However, long term benefits such as the establishment of small businesses and commercial agriculture have not been as forthcoming as expected by either the government or the locals. The people are reluctant to enter into further long term agreements with the government, such as leasing of land for plantation forests. The government underestimated the obstacles to development in the Gogol and did not follow through adequately with long term plans to be certain they would work.

Gogol people have been amazed at the scale and thoroughness of the logging. On paper the figures of the areas involved would have meant little to most people. The forest reserves are considered by many to be too small as sources of game and building materials. Some people had hoped animal populations would increase dramatically in reserves because of movement away from areas being logged. This has not occurred and large game such as wallabies, cassowaries, hornbills and goura pigeons have dropped in numbers throughout the Gogol Valley. The pressure on game in reserves has greatly increased with the dwindling area of unlogged forest and the greater mobility of hunters on the network of logging roads.

People's knowledge of wild birds and the extent to which birds are utilised is illustrated by the results of the questionnaire. Wild birds and their eggs are eaten fairly regularly and their plumes, which are highly valued, serve a major function in village ceremonies. People have mixed ideas about the effects of the logging on the birdlife but are generally concerned about the changes occurring. Their assessment of which birds are adversely affected (gauged by the same questionnaire) agreed well with studies I have conducted in the Gogol Valley. Given the opportunity local knowledge of the avifauna could be used extensively for research work in the Gogol.

Inhabitants in the Gogol have become bitter, believing they were deceived about the benefits and costs of the logging project. However, they see no alterantive than to pursue the path they have chosen and are therefore in a poor position to bargain for a better deal. From the beginning they had inadequate knowledge of the real worth of their assets and the likely consequences of any decision. People either imagined, or were encouraged to think of, a far brighter future than the one which has so far eventuated. If better communication had been established and more time spent to clarify the situation, many of the present difficulties with further development projects may have been avoided.

Much as we may desire to emphasize the human equation in decision-making about the use of their environment, it must be admitted that many examples of conflict have already shown (and will no doubt continue to show) that there is not just one homogeneous viewpoint by the local people. The reasons are many, and in any case viewpoints undergo rapid change in contemporary times, even in isolated areas of developing countries.

Case B: The Highlands of Papua New Guinea

In his cultural ecological studies, originating with small mammal surveys, in the highlands of Papua New Guinea, Dwyer has had cause to discuss the ecological focus of recent work by anthropologists, and the recurrent (but questionable) idea that 'resource utilization patterns are well matched to resource availability; that exploitation is not excessive; and that survival of natural resources is not jeopardized by traditional societal practices'.

He found many flaws in this idea once he lived with the PNG highlanders in the field, and concluded (Dwyer, 1981):

(1) That the theme of 'balance' or 'harmony', where this is expressed or implied, must be set against a growing awareness that Highland ecology has been changing in quite extraordinary ways throughout the past few thousand years. Cultural practices that appear to satisfy long term ecological goals of Highland societies are changing today and have been in flux for some thousands of years.
(2) That the theme of 'balance', expressed or implied, refers, in fact to environments that are far removed from 'natural'. Papua New Guineans have been busy modifying their environments for a long time. The anthropogenic grasslands of the Central Highlands are obvious examples, but, further west, the peoples of the great Papuan Plateau use rainforests in an extensive way such that below perhaps 1200 m. there is little forest that remains unmodified.
(3) Or again, the theme of 'balance', of living in harmony with nature, may be found in an attempt to reconstruct a past that, at the time of the study, no longer obtains.
(4) And finally, in most studies that I have read, hard data concerning wildlife resources is hard to find in the results presented. Instead we find extrapolations from a few cases of interpretations concerning wildlife that are founded in the theoretical model being espoused. While this might be the 'stuff' for academic theses it is not the information from which wise conservation decisions can be made by wildlife administrators.

Dwyer rapidly became aware that the secular and sacred of traditional societies are synergistically related; so that economics and ecology, hunting and good taboos, religion and social structure are inseparable. Thus in the climate of change following impact of Church and of Government, 'it is the intangibles that will wilt first'.

Of extreme relevance to a new kind of 'action research' is Dwyer's analysis of the 'conservation of sentiment', and the specific needs of legislation to ensure this.

The above discussion suggests several approaches and needs in the area of conservation legislation.

(1) The initial focus of wildlife legislation must be habitat preservation. This is necessitated by ignorance; we know little of the detailed habitat requirements of Papua New Guinea's wildlife. Rainforests in the middle altitude range merit particular attention. The retention, or promotion, of corridors of forest linking higher and lower altitude rainforests would be beneficial. In regions of high human population density where gardening land is in short supply it would be worthwhile exploring possible incentives to habitat preservation. For the relatively few species known to have very

restricted distributions the only approach is to cordon off the area concerned. Papua New Guinea could well adopt the principle that each of its species is sacrosanct—that none will be wittingly sent to extinction.

More enlightened approaches can only be based upon extensive base-line survey work. In Papua New Guinea this work can be easier than elsewhere for much of the requisite knowledge is stored by the people who live with the wildlife. It should not be too hard, I think, to train secondary and tertiary students in the art of scientific identification of at least birds and mammals and then to encourage those students to 'translate the known species of their *ples* into the terminology of science. At the very least, useful surveys based on inputs from schools could be conducted for a number of key wildlife species.

(2) Legislation should not be based upon theoretical statements concerning the utilization of wildlife except where these are sufficiently quantified. My own guess would be that the impact of trapping and hunting, *using traditional techniques*, is unlikely to be serious except where suitable wildlife habitat has already been greatly reduced. When return for effort declines there is less incentive to continue the chase. Wildlife will be jeopardized by human predation if suitable habitat is in short supply, if modern weaponry is used and if monetary rewards are high.

A special comment is in order about cassowaries. That modern phenomenon of the Mendi area, the 'cassowary race', is dramatic to the eye but may be hazardous to the birds. We need a measure of the threat to these grand creatures. If their survival is jeopardized—and they travel to the slaughter from as far away as the Nomad River—then legislation will surely be necessary. I will float one suggestion. If the price of cassowaries was fixed downwards, by the kilogram as meat, would the prestige of their owners fall and their own intrinsic value diminish in parallel?

(3) It is difficult to generalize in relation to the host of beliefs and customs found in Papua New Guinean societies that may contribute to wildlife conservation. Their contribution must be quantified before they can be adapted to legislation. I have offered a tentative framework within which their potential might be explored. Traditional practices that are found in religious beliefs or in prohibitions upon eating are probably less suited to modern conservation requirements than those found in land tenure. It is probably only those practices, already explicitly articulated in preservationist or management terms, that are amenable to modern needs.

Of course there are global considerations and responsibilities too. Rainforest is a fast vanishing resource, slow to regenerate and unable to sustain itself as small or greatly fragmented patches (Myers, 1979). Often it is dependent upon the fauna it supports, and the fauna of Papua New Guinea is very special; birdwing butterflies, pitted shelled turtle, green tree python, cassowaries, birds of paradise, bower birds, salvador's teal, tree kangaroos, long-nosed echidnas, the remarkable earless water rat. All these are absent or poorly represented outside New Guinea. They and many other species should not be lost and zoos can never be enough. Most will need large reserves to guarantee survival; even larger reserves if they are to continue to provide meat and ornamentation.

But legislation that is modelled on the above considerations seeks to conserve wildlife in an anonymous sense. It is distanced from the people whom hopefully it is intended to serve. Its compatibility with traditional practices may be largely beside the point once those practices are removed from their traditional setting. In a country that is consciously altering relationships between people and nature the conservation of wildlife must ultimately be dependent upon public attitudes.

Conservation must be a self-conscious and a public decision; it cannot be achieved in anonymity. It is essential, therefore, that appropriate public attitudes are fostered, that the emotional attachment of people to wildlife is sustained.

In Papua New Guinea there is enormous scope in this area. For millennia Papua New Guineans have found animals 'good to eat'. They have also found them 'good to think'.

Case C: Cape York Peninsula, Northern Australia

Work in Cape York Peninsula over the past 2 years has led the ecologist Dermot Smyth, of the Australian Institute of Aboriginal Studies, to a disconcerting realization that the researcher may be doomed to remain an 'outsider' and even an alien, and that (as De'Ath found) any land assessment technique is inevitably bound to value-judgments. In a personal communication, Smyth writes:

A study of current land assessment procedures reveals at least one common denominator: land, at whatever scale, is viewed from the *outside* by a scientific *observer*, or team of *observers* concentrating on particular environmental factors (soil, climate, vegetation etc.) and then integrating these data to suit the chosen assessment system (biogeoclimatic, biophysical etc.).

Even when personally engaged in field studies (as distinct from air photo or Landsat interpretation, soil analysis, computer analysis etc.), the scientist maintains his *observer* status. That is, he comes to observe and assess the environment and does not claim to be a *participant* therein. This leads to notions of primary, secondary, tertiary classification etc., and to a belief that the observer-scientist's report can represent an original assessment of 'virgin' lands.

Given a generally overriding emphasis on exploitation potential, many existing assessment systems contain only grudging reference to the existing, non-productive biota. Thus even when human occupation by 'primitive forest dwellers' is acknowledged, their activities (hunting, gathering, shifting cultivation etc.) are placed on the exploitation side of the assessment equation and quite separate from the assessment procedures.

In summary, established land assessment procedures involve a scientific examination of selected components of a land ecosystem which, when assembled in certain ways, are expected to be of predictive value for the 'efficient, safe and pleasant' use of that land for development according to Western styles.

An antithesis of this approach is the empiricism of the original human invaders of the forest lands. As hunters and gatherers, or subsequently as shifting cultivators, the original people had a very personal interest in land assessment and exploitation prediction. That their methods were successful is evidenced by the continuance of these forms of subsistence in modern times. Three important factors militated in favour of success. Firstly, population densities were low; secondly there was a willingness (albeit an adaptive necessity) to alter the prevailing human culture to match the limitations of the land; thirdly, and as a consequence of both the foregoing, there was sufficient time (many thousands of years) for the empirical approach to be effective.

This is not to say that the history of hunter–gatherer and then shifting cultivator occupation of forests and other lands is one of total harmony with, and minimal transformation of, the environment. Indeed the contrary is clearly the case in many areas e.g. the grasslands of the New Guinea highlands. More relevant, however, is to understand that this history is one of human *participation* in the environment and of cultural accumulation of detailed and subtle ecological appreciation.

The significance of this history to land assessment procedures is two-fold. Firstly, modern societies, though usually absentee exploiters of forest lands, are not free from many of the constraints which shaped the hunter—gatherer societies. The need to curb population growth, for example, is more pressing now than ever before. In more general terms, there is a need to explore the possibilities of adapting modern societies to match the produce (nutritional, technological, aesthetic) of the land, rather than always insisting that the land be manipulated to suit the demands of society.

The second major significance of the hunter—gatherer experience in relation to modern land assessment is simply to recognize that these societies were and are a great source of ecological knowledge. That is, their experience, empirically gained and culturally maintained, should contribute to the assessment side of the equation just as their demands as hunters, gatherers and cultivators are considered among the 'options for exploitation'.

Case D: Arnhem Land, Northern Australia

My reference here is to the work of Haynes, a forester with strong biological and anthropoligical interests, working in Aboriginal communities in northern Australia. The outcome of a European-conceived and directed forestry operation (logging, tending, and planting) in Arnhem Land, Northern Territory, provides a salutary case history (Haynes, 1977).

Well aware of the difficulties for Europeans in describing and interpreting the feelings and aspirations of Aboriginal people, Haynes—with the assistance of Aboriginal informants and anthropoligists—began to analyse the specific relationships of the Aborigines with the land at Maningrida. He (as Dwyer did) noted that not only was it impossible to separate secular and sacred relationships, but also that 'the more sacred the matter, the more private it becomes'. An abject failure of communication is revealed by his statement that 'after more than 14 years there was not to be found *one single Aboriginal person* in the area who understood what forestry was really about'. Haynes was eventually invited by the Maningrida Council to recommend an alternative Aboriginal-style forestry operation. The new 'indigenous' forestry developed rapidly, and Haynes wrote:

> The Council itself, and its employees, now performs some of the functions previously undertaken by the Government. Some logging and saw-milling are being run to supply local needs for timber. Thus Klye-Little's dream of Aboriginal people doing the job by and for themselves has at last become a reality. European assistance is confined to the part-time efforts of one man.
>
> Moreover, important differences of style are emerging. Land owners are doing the conservative work in their 'old custom way' which involved burning off the land during the early part of the winter 'dry' season. The details of how this is done should be of interest to managers of all Australian native forests. In addition, some groups of people living on 'outstations' are building bridges and tracks with, again, very limited assistance. Perhaps most important of all, some Aboriginal groups are expressing their dreams for the future; dreams that are bound up with developing *their* forests in ways that are compatible with their lifestyles.

There is, finally, a lesson to be learned from the whole sad story. It is simply that large organizations cannot afford to be aloof in pursuit of objectives which are not in tune with what is happening in the community. At Maningrida a few hundred unsophisticated people were able to take on and defeat a comparatively large organization. It seems to this writer that we have all become over impressed with our technology, and absorbed by it; so that in the end events have overtaken us.

CONCLUSIONS

It is argued in this account that scientific research into forest management and its socio-economic aspects should originate among the local people whose lifestyles and welfare are most affected by the proposed logging operation or other forestry procedures. Such research should initially and throughout the project involve consultation with and participation of local village people, among others, well before the forest is logged. The local people should be able to employ their own advocates and experts to assist their deliberations.

Results of research should be relayed to the people throughout the project and their reactions and further suggestions used if necessary to modify the research programme. Special attention should be paid to the concern of local people with less tangible values of the forest and the 'conservation of sentiment'.

Conventional land-use planning based on western-style techniques and imposed from 'top-to-bottom' should be replaced by 'bottom-to-top' decisions based on the realities of the indigenous people's lives. A balance will then need to be struck with a new and flexible flow of information and decision in both directions to co-ordinate development in the best interests of the local people.

Legislative and institutional changes to implement research recommendations for forest management should be compatible with traditional practices, and geared to fostering appropriate attitudes among the community and as a whole throughout the village, the region and the nation.

The extracts from the recent conclusions of field workers in cultural ecology and social anthropology in Papua New Guinea and northern Australia, given here, illustrate the research problems, and the rationale for a new, people-oriented methodology that is urgently required.

REFERENCES

Anon (1976). *Report of Symposium on Ecological Effects of Increasing Human Activities on Tropical and Subtropical Forests Ecosystems.* University of Papua New Guinea, 28 April–1 May 1975. MAB Publication No. 3, Australian Govt. Publishing Service, Canberra.

Anon (1977). Papua New Guinea participation in MAB. *Nature and Resources*, **XIII,** 24–50 (UNESCO).

De'Ath, C. E. (1978a). 'Learning to Live with Chainsaws and Jinkers: an Involved Social Critic Looks at Research being done on the Effects of the Gogol Timber

Project'. Paper presented at *Man and the Biosphere Workshop*, Madang, 17–20 March.

De'Ath, C. E. (1978b). *Pots, Power and Progress in the Trans-Gogol of Papua New Guinea.* The Second International Symposium on the Art of Oceania. Victoria, University of Wellington, Wellington, New Zealand.

De'Ath, C. E. (1980). *The Throwaway People: Social Impact of the Gogol Timber Project, Madang Province.* Monograph 13, Institute of Applied Social and Economic Research, PO Box 5854, Boroko, Papua New Guinea.

Dwyer, P. D. (1981). Wildlife conservation and tradition in the New Guinea Highlands. In *Monograph* by *Institute for Applied Social and Economic Research*, Waigani, PNG. (In press).

Haynes, C. D. (1977). Land, trees and man (Gunret, Gundulk, Dja Bining). Maningrida Council, Maningrida, NT, Australia. *Commonwealth For. Rev.*, **57,** 99–106.

Jonas, W. (1976). 'The forest ecosystem and the development process in less developed countries'. In Report of *Symposium on Ecological Effects of Increasing Human Activities on Tropical and Subtropical Forest Ecosystems.* MAB Publication No. 3, pp. 159–71. Aust. Govt. Publ. Service, Canberra.

Lamb, D. (1977). Conservation and management of tropical rain-forest: a dilemma of development in Papua New Guinea'. *Environmental Conservation*, **4** (2), 121–9.

Myers, N. (1979). Islands of conservation. *New Scientist*, August, pp. 600–2.

Qureshi, A. H. *et al.* (1980). *Assessing Tropical Forest Lands, their Suitability for Sustainable Uses.* East–West Environment and Policy Institute, East–West Centre, 1777 East–West Road, Honolulu, Hawaii-96848.

Samol, G. D., and Webb, L. J. (1978). Report on Papua-New Guinea, *Trans. Second International MAB—IUFRO Workshop*, pp. 21–5, October.

Seddon, G. (1978). Lessons of the Gogol. In Viewpoint, *Newsletter*, No. 6, Macquarie University, Sydney, NSW, Australia.

Webb, L. J. (1977). *Ecological Considerations and Safeguards in the Modern Use of Tropical Lowland Rain Forests as a Source of Pulpwood: Example, the Madang Area PNG.* Office of Environment and Conservation, Dept. of Natural Resources, Waigani, Papua New Guinea.

Socio-economic Effects and Constraints in Tropical Forest Management
Edited by E. G. Hallsworth

16

Native Americans and the conservation of flora and fauna in Brazil[1]

ANTHONY SEEGER
Museu Nacional, Rio de Janeiro, Brazil

INTRODUCTION

The success of conservationist measures to protect the flora and fauna of a given region inevitably rests upon the role of the local human populations in respecting and cooperating with the decreed measures. This respect comes not only from education, but also from the provision for adequate subsistence independent of the wild flora and fauna. In addition to implicit consent, there must be enforceable sanctions against abuse of the laws, and no doubt about the legitimacy of the control of a given area, or over the exploitation of a given species or natural resource. For a number of reasons, which have nothing to do with the conservationist legislation presently adopted in Brazil, such tacit consent and unchallenged legitimacy are not everywhere present. One problem is the remaining dependency of the local populations on the forests for their basic protein supplies; another is the conflict between conservation laws and legislation which protects the surviving native American, or Indian, societies.[2] The Indians, specifically, are exempt from all laws restricting the hunting and fishing of endangered species.

This paper examines some of the difficulties which arise when native American reserves and forest reserves, ecological stations, or national parks are juxtaposed, or when a single area is expected to serve for both ends. Three cases are examined here: the conflict between the Pataxó Indians and the Brazilian Institute for Forest Development (IBDF) with respect to the Mount Pachoal National Park in the State of Bahia; the multiple use of the Ilha de Bananal, site of the Araguaia National Park in the State of Goiás; and the proposals for an Indian Park in the northern part of the Territory of

Roraima for the Yanoama Indians, who reside in part of the Nublina National Park. The most important conclusions to be drawn from the situations described are: (1) that human populations adapt rapidly to new technology and to the pressures of the market economy, and change their relationship to their ecosystems far more rapidly and drastically than the animal populations; (2) separate provision must be made to guarantee the use-rights of native American areas, as guaranteed by existing national law; (3) the native Americans should be provided with technical assistance and advice to prevent the destruction of their natural resources; (4) areas designated for conservation and protection should be very clearly distinguished from Indian areas in their establishment, their maintenance, and their politics; and finally (5) the non-Indian regional population must be provided with substitute resources and the areas must be adequately policed.

The Office of Environmental Affairs of the World Bank has developed a definition of what they call 'tribal groups' which attempts to distinguish such groups from 'peasants' and other minorities. The definition is worth quoting in full, as it applies to many, but not all, of the human populations under study here:

> The term 'tribal people' often shortened to tribal or tribe, is employed here to characterize a specific type of population. Such groups typically have stable, low-energy, sustained-yield economic systems. More specifically, the people may be hunter-gatherers, shifting agriculturalists, herders, simple farmers or fisherfolk. The populations included in this terminology exhibit many, if not most, of the following characteristics:
>
> (a) geographical isolation or semi-isolation;
> (b) unacculturated or only partially acculturated into the national society;
> (c) non-literate; not possessing a written language;
> (d) non-monied, or only partially monetized; largely or entirely independent of the national economic society;
> (e) ethnic distinctiveness from the national society;
> (f) linguistic difference from the national society;
> (g) possessed of a common territory;
> (h) economic base more tightly dependent on their specific environment;
> (i) possessing leadership, but no national representation, and few, if any, political rights.
>
> (World Bank, 1981: 5)

The report suggests that special provisions be made for such societies when they are encountered in areas which are being targeted for national or regional development and for which World Bank loans are solicited. Since their economic system is not energy intensive, and depends upon an intimate relationship with a given ecosystem, the Brazilian Indians require a fairly large geographical area. Any reduction of alteration of the ecosystem will have drastic effects on the groups involved.

In their dependence on the ecosystem, the tribal societies resemble the flora and fauna of the region, and it is significant that the report was written within the Office of Environmental Affairs. Such an attitude, although important for the protection of the tribal societies, has often led to the superposition of ecological reserves and Indian areas. Although the initial relationship between the Indians and the ecosystems resembles those of the other species, that is a temporary state of affairs. A misconception about human societies can cause serious problems.

TRIBAL SOCIETIES IN BRAZIL

There are approximately 140 Brazilian Indian societies, speaking many different languages, and found unevenly distributed among all the states, with more than half of them found in the Amazon region. The size of the groups is fairly small, ranging from communities of a few individuals to loosely designated 'tribes' (communities in contiguous areas that speak a mutually intelligible language) of eight to ten thousand. The total Indian population in Brazil is not thought to exceed 200,000. Under the Brazilian constitution the Indians are guaranteed permanent possession and exclusive usufruct of the natural riches and the useful resources existing on their traditional territory (Article 198).[3] They have a national guardian, in the form of the National Indian Foundation (FUNAI), whose ends are, among others, 'to assist and protect the forestdwellers, their culture, and their patrimony: to guarantee to the Indian the unalienability and the permanent possession of the lands they inhabit' (Presidência da Republica, 1981:1).

The Brazilian Indians, whose diversity makes them more difficult to assist and protect than their small numbers would indicate, are commonly divided into four categories according to the degree of their contact with the national society, and particularly their integration into the regional economy and their dependence on manufactured goods. These categories generally follow a scheme elaborated by Darcy Ribeiro (1957, 1977), dividing the groups into *Isolated, Intermittant Contact, Permanent Contact*, and *Integrated*. (The Indian Statute reduces these to three: Isolated, In the Process of Integration, and Integrated; the World Bank uses the four classifications.) This division overlooks important differences among the groups, and may give the false impression that it is the fate of all societies to start out 'isolated' and end up 'integrated'. In fact most Brazilian Indian societies have been decimated, become extinct, or been forced against their will to 'integrate' into Brazilian society rather than follow the ideal pattern described below. Even so, the categories are useful heuristic devices if used with care.

Isolated groups are those that are completely uncontacted, or which rarely enter into contact with members of the national society. They are virtually autonomous and are characterized by self-sufficiency. Groups in *intermittant contact* are those whose contact with the national society is restricted to particular representatives such as traders, missionaries, or government functionaries. These groups have already begun to establish economic relations through barter for a growing number of 'necessities' as perceived by them: steel tools (often obtained before contact by barter with other Indian groups or by raiding), fishline, fishooks, trade beads, etc. At this point the groups are particularly susceptible to diseases, and to derangement caused by moves to different areas with different ecosystems. *Permanent contact* describes tribal groups maintaining regular, permanent communication with different representatives of the national society. There is a growing dependence upon manufactured goods and this leads to increasing participation in the national economy in order to obtain them. Here problems of health recede before problems of exploitation of Indian labour and expropriation or illegal invasion of their lands. *Integrated* groups are considered to be virtually indistinguishable from the rest of the rural population, although they identify themselves as a tribal population and are so identified by their neighbours (see World Bank, 1981: 7).

One of the things to remember when using these classifications is that a given group can be transformed in a number of ways. A group may return to isolation after intermittent contact; the degree of its dependence on the national economy may well vary according to changes in the regional economy. Some societies become extinct before they are contacted by the members of the national society—many were massacred by other Indians, or died from diseases transmitted by them, before ever seeing a non-Indian. One of the most notable developments of the past few years in Brazilian Indian affairs has been an increasing awareness on the part of the Indians of the threats to their survival, and their attempts to negotiate a different relationship with segments of the national society. These efforts may be seen in the renewed hostility between Indians and non-Indian settlers in the central west, and also in the attempts to gain a greater control over the profits from the resources on their lands as among the Gaviões in Pará, the Pareci in Mato Grosso, the Kaxinaua in Acre, and other cases.

Isolated tribal societies in Brazil are generally thought to have established an 'equilibrium' with, and an optinal adaptation to, their ecosystem (see for example Carvalho, 1981: 33; Meggers, 1971). Such ideas are part of the 'common sense' of the man-in-the-street in Brazil. Although they may have had some important effects on the ecosystems over the long term, it is obvious that their capacity for drastically altering it is limited by their technology. But the societies themselves vary widely in their own attitudes toward the preservation of a balance between men and animals or 'nature' in general. Some, such as the Gê-speaking societies in Central Brazil (Pará, Mato Grosso, Goiás and Maranhão), have no evident concern for conservation. They happily kill as many animals as they can, exploit the local resources, and move their villages frequently. During the last five hundred years they have expanded their territory considerably, occupying the areas of other societies (Seeger, 1981). Other societies, such as those in the Northwest Amazon (Upper Rio Negro), are extremely concerned with the maintenance of relationships both with natural species and with other human groups, through contention and restrictions on hunting (Reichel-Dolmatoff, 1976). These communities have not moved much during the last few hundred years, and seem to establish fairly complex relationships with both riverine and forest ecosystems, as well as with other Indian groups (Goldman, 1963; Wright, 1980). Between the high mobility of the Gê and the sedentary characteristics of the Tukanoan societies, with their respective attitudes toward natural species, may be found many other Brazilian Indian societies. Thus it is incorrect to say that the Indians are 'naturally' conservationists. Although the Brazilian Indians may not have made profound alterations in their ecosystems before contact, this does not always mean they had a cosmological commitment to conservation; often it is simply because they had no choice.

In spite of these real differences among the groups, in the eyes of the Europeans the native americans were 'nature people' (*Naturvolken*) or 'beings of the forest' (*silvícolas*) whose lifestyle was 'natural'. That is still an idea very present in many conservationist movements in Brazil and abroad. Seen from this perspective, it is quite appropriate that Indian reserves and areas devoted to protecting natural species should be established jointly.

When attempting to establish Indian reserves or conservation areas, it is important to establish them before extensive development of the region, with its accompanying deforestation, depletion of natural species, and massacre or infection of unprotected Indian communities by disease. Under these conditions, the Indians' relationship with their own ecosystem is usually little altered. It is especially useful to ally conservation with Indians because the political force is greater when the two groups make common cause (and it is consequently seriously affected with the two groups fighting among themselves). But it is important not to overlook the differences between native and animal populations.

The prime difference between human societies and natural species is that human societies are far more adaptable. The Brazilian Indians, once they have made contact with segments of the national society, have inevitably changed their relationship with their ecosystem. Even if they retain their traditional lands (which is certainly not always the case), they begin to exploit them more intensively and with greater selectivity. This change in their relationship is the direct result of their entry into the regional economy, to their peculiar legal situation in Brazil, as well as to the activities of the Brazilian Indian Foundation. In many cases the Indians have been used as agents for the depredation of natural resources or the decimation of natural species by non-Indians. In other situations, however, access to the resources necessary to the Indians' own physical survival has been denied to them in the name of conservation. The political and economic autonomy of many groups and their access to their traditional lands and resources, in certain cases come dramatically into conflict with the elementary requirements of adequate conservation.

CHANGING RELATIONSHIPS BETWEEN NATIVE AMERICANS AND THEIR ECOSYSTEMS

By Brazilian law, the Indians are given the judicial status of 'relatively minor', which characterizes children between the ages of 16 and 18, the infirm, the mentally ill, and a few other categories. The National Indian Foundation is the government appointed guardian. This status for the Indians, in spite of its ambiguity, brings with it a number of benefits: collective ownership of land, less-than-full responsibility (for crimes which may arise out of misunderstanding rather than criminal intent), and protection from some of the most flagrant abuses seen in other countries at other times—such as the exchanging of large areas of land for small amounts of liquor (as happened repeatedly in the United States). The success of the law depends on the ability and sagacity of the guardian, and its dedication to acting in the long-term interests of its wards. Sagacity and loyalty to its wards have not always characterized the actions of the National Indian Foundation, which is poorly accommodated within the Ministry of Interior, many of whose overall policies are not in the Indians' interests.

One of the aspects of the special status of native American communities in Brazil is that they are not ruled by Brazilian laws regarding the protection of endangered species, and the hunting, fishing and exploitation of natural resources.

The Indian immunity has meant that they are often employed as professional hunters. They kill for pelts of endangered species, or engage in commercial hunting and fishing where it is prohibited, and sell their catch to smugglers who make large amounts of money on the transaction. The contraband is widespread including, in addition to pelts and flesh, pet birds and mammals, orchids, etc. (see Carvalho, 1981).

Another cause of the depredation of natural species is the commercialization of Indian artefacts as tourist art. Indian artefacts have a wide market within the cities for both Brazilian national and foreign buyers. Almost universally, the agencies which work with the Indians commercialize their artefacts: missionaries, local traders, and the Indian Foundation. The Indians themselves may also sell artefacts along highways in the more densely populated parts of the country. The most intense commercial development of Indian artefacts has been undertaken by the Brazilian Indian Foundation whose *ARTEINDIA* stores are found in most major airports, bus stations and cities. In addition to the deleterious effects on the native crafts, the traffic in artefacts may lead to the exhaustion of the natural raw materials. Instead of fabricating twenty parrot or macaw feather headdresses in a year, a group may start to produce two hundred or more, with resulting pressure on the parrot population. Although feather art is the most obvious difficulty—domestic species do not have the same colourful plumage, even when their feathers have been dyed—even such apparently innocuous crafts as basketry may lead to the disappearance of a given plant species in a large area. Although apparently a good way of exchanging manufactured goods for a marketable product, native artesanry gives official sanction to further depredation unless it is undertaken with great care.

The Indian societies, which desire the manufactured goods that make their lives easier (at first) and then as they adopt non-Indian attitudes, believed by them to be more 'civilized', are stimulated to alter their relationships with their ecosystem and to exploit it with greater intensity. But that is not the full extent of the alterations. The National Indian Foundation, and its predecessor the Society for the Protection of the Indians (SPI) have initiated large projects on a number of Indian areas: rice, soybean, and coffee plantations, Brazil-nuts, cattle raising and lumbering have been among the largest of the many activities engaged in. The rationale is that the Indian lands should produce enough revenue to pay most of the expenses of the maintenance of the Indians throughout the country. The general effect is that the land and sometimes the labour of a given society is used, but the money does not return to the Indians who are guaranteed the usufruct of their holdings. It goes to Brasilia where it is incorporated into the Foundation's budget and pays for other projects. In some parts of the country, especially the South, the exploitation of the Indian reserves by the agency that should be looking after their interests, as well as unchecked illegal commerce, has resulted in the denuding of Indian forests and the loss of irretrievable patrimony (Simonian, 1981). When the processes described above result in the scarcity of needed natural resources on the Indian reserves, the Indians will put increasing pressure on any conservation areas nearby.

Finally, even after many years of contact, the Indian communities are largely dependent on wild protein for subsistence. They share this dependence with most of the peasant population of the Brazilian interior, whose principle animal protein is derived from fish and game. Indian lands are often invaded by non-Indians for hunting, and the resources of reduced areas will prove to be inadequate. Some Indian lands are also inadequate for agriculture; they may be forced to use conservation areas for subsistence agriculture and hunting or fishing.

Below I will describe briefly three quite different cases which illustrate some of moral difficulties and legal intricacies of the relationships between conservation and Indian rights. These are the case of the Pataxó of Porto

Seguro, the Karajá of the Ilha de Bananal, and the Yanoama of Roraima. They all have in common the juxtaposition of national parks and Indian areas. The things that vary are temporary: the relative size of the Indian and conservation areas, the density of the regional non-Indian populations, and the means and ends of resource use on the part of the Indians. All of these are likely to change over time.

The Pataxó: struggle over a symbol and survival

The Atlantic Forest which once covered the coastal regions of Brazil, has almost completely disappeared. Clearing for agriculture and intensive lumbering reduced it to small areas in Bahia and Santa Catarina. One of the isolated remainders is in the municipality of Porto Seguro, in the south of the State of Bahia. Faced with the extremely rapid deforestation in the 1950s, the Mount Pachoal National Park was established by decree in 1960. The area of the Park was then inhabited by both Indian and non-Indian populations. Their holdings were expropriated, but only the non-Indians were actually removed. A group of Pataxó remained within the confines of the Park (Agostinho da Silva, 1978).[4]

The presence of the Indian community on National Park lands led to conflict between the National Indian Foundation and the National Institute for Forest Development (IBDF) which is responsible not only for the protection of flora, but for the protection of fauna as well (by separate acts: law number 4771 of 15 September 1965 on forests, and law number 5197 of 3 January 1967 on the protection of fauna).

The Brazilian Indian Foundation began to take measures to turn the area occupied by the Pataxó into Indian lands in the 1970s. In response, the IBDF began to persecute the Indians, prohibiting both subsistence and market-oriented economic activities within the National Park. The Indians were warned off the Park and their small garden plots were destroyed when discovered (Agostinho da Silva, 1978:2). They were reduced to collecting crabs on the coastal waterways and to manufacturing and selling artefacts. In spite of this, the Indian population increased during the decade to over 700, as dispersed Pataxó moved to Porto Seguro in the hopes of obtaining some land.

The lands occupied by the Pataxó had been designated as Indian lands in 1861, and occupied by them at least since 1892, according to a census in that year. Their legal rights to continued occupation of the area are quite clear; equally clear is the imminent extinction of the Atlantic Forest.

A team of anthropologists from the University of Bahia studied the Pataxó and suggested maintaining the group in its present location and made a calculation of the land area required by the group—an area which assured them swidden plots with rests of 30 years, giving a total of 2197.8 hectares. Then they calculated a 3 per cent growth per year for 23.5 years (approximately a generation) and a 20 per cent growth through in-migration of relatives, and came up with the figure of 5300 hectares of necessarily rich agricultural land for the group. This suggestion, carefully developed using cultural and economic criteria, was ignored. It was considered too large an area of the remaining forest. Instead the Pataxó were alloted lands that are unsuitable for agriculture, and are once again reduced to hunting for crabs and selling artefacts. The population is declining as the community disperses in a search for adequate living conditions.

The impasse in Porto Seguro is the result of the injustice on both sides. The area that remained was inadequate to meet the Indian needs and still preserve the Atlantic Forest. The issue is quite clear. The National Park was established too late, when little option remained except for lands already occupied by Indians for over 90 years. The total amount of land area included was too small to share with the Indian population which could be expected to grow. Two alternatives would have been possible: in 1960 to have reserved a larger area for the National Park, assuring the rights of the Indians, or to give the Pataxó an ample reserve in the rich agricultural areas to the North. The social and economic costs of disappropriation would be extremely high, and constant vigilance would be necessary to protect the Indian lands from invasion. The solution reached was the worst possible for the Pataxó. They were allotted lands totally inadequate for subsistence agriculture.

The Karajá, agents of depredation

Bananal Island, a large riverine island which lies between the states of Goiás and Mato Grosso, is formed by two branches of the Araguaia river. Early explorers of the river encountered the Karajá Indians resident on the island and in adjacent areas. Today, the island has a National Park with a total of 563,312 hectares in the north, under the jurisdiction of the IBDF (although there is also a Karajá settlement on the northern tip). The central part is an Indian reserve, in which are found most of the Karajá tribe, and in the South there are large cattle ranches. All parts of the Island, including Indian and National Park lands have been illegally invaded by cattle ranchers. Some estimates put the number of illegal ranches at over one hundred (José Candido de Melo Carvalho, personal communication).

The Indians are principally, and traditionally, a riverine population. With a long and difficult history of contact, the Karajá still maintain many of their traditional customs and subsist on the products of their gardens and rivers, as well as the income from their extensive sale of artisanry and illegal commerce in natural products. The contact situation in the region is quite complex; the National Indian Foundation (FUNAI) has invested considerable quantities of money and personnel in the area, and yet the difficulties persist.

Since the non-Indian populations are prohibited from hunting the protected species that are still to be found in the region, such as the ararinha (*Pteronura brasiliensis*) the jaguar (*Leo onça*), and the jacaré (*Melanosuchus niger*), the Indians are paid to hunt them. Since commercial fishing and hunting are prohibited, the Indians are encouraged to fish and hunt for turtles for resale (Dante Teixera, personal communications). The Indians are not the only ones to exploit the legally protected species, but their legal protection and subsequent exploitation is quite serious in its long-range implications.

The Karajá are among the most prolific suppliers of Indian artefacts in the nation. They are particularly famous for their clay anthropomorphic figures (Fenelon Costa, 1968), but they also produce a number of other 'Indian' items such as small bows and arrows, certain feather ornaments, wood carvings, and so forth. The raw materials for some of these require large-scale predatory expeditions. In 1980 a group of Karajá men on collective hunt for the Colhereiro (*Ajaia ajaja*) whose feathers are used in artefacts, resulted in a kill of between 250 and 300 individuals (Dante Teixeira, personal communication).

There can be no doubt as to the legitimacy of the Indian claims to the resources of the Island. The situation is not as desperate as that of the Pataxó in Porto Seguro, because the population density is less, the Indians have lands of their own—on which the National Indian Foundation is raising cattle for sale and the regional population is far less. One cannot neglect the seriousness of the illegal occupation of large parts of the island and parts of the Park by non-Indian ranchers, however. The difficulty on Bananal is that the Indians are often used as agents of middlemen who supply pelts, fish and pets to the national and international markets, thus altering in fundamental ways their relationship with the ecosystem, and affecting their own patrimony as well as that of the nation.

The Yanoama: the search for an adequate solution

The Yanoama are Brazil's last large group of isolated tribal communities. Located on the Venezuelan border in the Territory of Roraima, the Yanoama are in fact five dialect groups with a population of perhaps 8300 on the Brazilian side of the border, living in something over 200 widely scattered communities. These communities maintain complex relationships of intermarriage, alliance, and feuding among themselves (Ramos and Taylor, 1979). Their exploitation of the ecosystem is quite traditional: they do not supply regional markets and in fact they have only sporadic relationships with representatives of the national society.

During the last five years there has been an international campaign to provide adequately for the future of this large and vulnerable group. This campaign, begun in Brazil, seeks to assure them adequate lands, protect their health, and control the development in the northern part of Roraima. Since intermarriage and freedom of movement are important for the groups, it has been argued that a single, large, continuous reserve (formally a *Parque Indígena*), would be the only way to meet the needs of the groups and to honour the constitutional guarantee. The suggestion, fully described in *The Yanoama in Brazil 1979* (Ramos and Taylor, 1979) suggests a Park with a total area 6,446,200 hectares. This includes 3,270,800 hectares of the area which are covered by environmental protection laws by virtue of their altitude, or their low natural capacity.

A total of 74.14 per cent show the characteristics of an environmental protection area (Ramos and Taylor, 1979: 138–139). The National Indian Foundation made a counter move, and decreed 21 areas ranging from 10,900 hectares to 442,500 hectares, with a total of 2,228,270 hectares. The difficulties of these small areas are the facility with which they could be invaded by non-Indians, and the impossibility of the free migration of the groups. Furthermore, moving groups around to place them in small reserves is extremely delicate, due to their dependence on rather specific ecosystems. (Ramos and Taylor, 1979: 141). The authors suggest that

> The area of the proposed Yanoama Park is particularly suitable not only for the preservation of an ecologically noteworthy region, but also for the protection of the last large nation of Brazilian Indians, with no detriment to regional development. (Ramos and Taylor, 1979: 136.)

There has been an impasse over the Yanoama case, which awaits a final decision. The search for this solution has resulted in a compromise suggestion for a multiple-use area—one that would allocate certain regions as Indian areas, others as ecological areas, forest reserves, and national parks, and yet others as areas under military control.

At first sight a multiple-use area would seem to be an ideal solution for the Yanoama. They could continue to live where they are, and they would be protected from many of the ill-effects certain to come from the establishment of many mini-reserves through the federal control of the lands between Indian areas. But is it an ideal solution? Viewed in the light of the cases we have already discussed, there are a number of conflicts that can be predicted. The Yanoama population, it is to be hoped, will increase over time; their use of their areas will inevitably change due to their integration in a regional, national, or international economy. How will the flora and fauna be protected? How will their own futures be protected?

CONCLUSIONS

The preservation of natural eco-systems and species, and the protection of tribal people are, at one level, parallel endeavours. Both must be provided for at the earliest possible instance, and both must be protected within the parameters of the national political arena. Since the Indians are perceived, somewhat romantically, to be part of the 'natural' ecosystem, there has been a tendency to superimpose Indian lands and conservation-designated areas. This has been done by multiple usage, by creating national parks within Indian areas, or by the simple existence of Indians in national parks. But native American societies alter themselves far more than the animal populations, and attempts to provide for both in the same, or in limited adjacent territories, are often based on erroneous logic, or no logic at all. The political possibilities of a united front consisting of both conservationists and Indian rights groups makes a joint effort for establishing multiple-use areas quite reasonable, and it also makes disagreements within the ranks so much more serious.

This paper has argued that Indian societies do not necessarily maintain a single relationship with their ecosystems. They are capable of considerable transformation and long-term survival, as is the case of the Pataxó who have had hundreds of years of contact, and yet who would still rather re-group on land of their own rather than disperse into the local population. The examples cited also show the Indian populations are used by non-Indians to exploit natural resources protected by law. It has been shown that where the resources are limited, the confrontation between Indians and forest management has no real solution. In Brazil, however, there *are* large tracts of land which are not occupied by indigenous inhabitants. Therefore it is possible to reach solutions which satisfy both parties, if land is set aside before the regional population grows. Instead of conflicting over multiple-use areas, conservationists, Indians, and Indian rights groups will have to join together for a separation of areas in such a way as to guarantee the national patrimony and the tribal rights.

In the case of the Yanoama Park proposal, the areas of competence will have to be defined with extreme clarity, and provisions made for educating both Indians and non-Indians about the usage of the different areas. It is a solution for immediate problems, but it will raise new ones which are best solved now. What will be the Indians' rights in the non-Indian part of the Park? Which areas will serve for what ends? How will this be decided? When can the decisions be renegotiated? What appeals will be granted to the Indians? All of these are questions which should be worked out as soon as the solution is found—if it is a multiple-use kind of solution.

This paper has discussed the difficulties arising where Indian lands and National Parks, or conservation-oriented areas are superimposed or connected. But native populations are certainly not the only threat to the survival of natural species. The lack of adequate policing, with the resulting extensive contraband of pelts, timber, meat, pets, and other resources, is also an important factor. Even more important, the local populations in Brazil must have an alternative to wild animal protein accessible to them. Carvalho (1981)

Table 1 Meat prices in suburban Manaus (Carvarho, 1981)

Mammal	Price per kilogram of meat
Capybara (*Hydroachaeris hydrochaeris*)	US$2.00
Freshwater porpoise (*Trichechus inunguis*)	US$1.35
Wild pig (*Tayassu pecari*)	US$1.35
Deer (var. species) (*Mazama* spp.)	US$1.35
Armadillo (*Dasypus* spp.)	US$1.00
Domesticated chicken	US$3.00
Domesticated pig	US$2.00
Beef	US$5.85

cites the prices of the flesh of a few wild and domesticated animals in suburban markets in the city of Manaus (Amazonas) which I think speak for themselves (I have converted the prices he gives into dollars at the September 1979 rate of exchange—the time when the data were collected). These are listed in Table 1. Since the minimum monthly wage at that time was approximately US$110.00, it can be seen that the local populations cannot afford *not* to eat wild species. Some provisions will have to be made to change this dependence on the natural ecosystem by an ever increasing population.

The survival of the flora and fauna depend on massive appeals to public opinion and also to alterations in the material conditions of life in the region. The survival of the Brazilian Indian societies depends upon the permanent guarantee of their lands, continued health and education assistance, and increasingly on technical advice so that they do not repeat the mistakes made by sectors of the national society on their own patrimony. Special attention should be given to the deleterious social and ecological effects of artefact production and large scale ranching, as well as to the possibility of substituting raw materials and planting or raising the sources of the materials used. The Brazilian Indian Foundation must be especially careful not to destroy the patrimony it is entrusted with preserving by ill-advised large scale projects on Indian lands.

The issues of conservationism and of Indian policy are quite distinct. They should not be confused. The price paid eventually for mistakes at the start is a very high one, in both human and ecological terms.

NOTES

1. I would like to thank my colleagues at the Museu Nacional who read earlier versions of this paper or discussed the ideas with me, especially José Cândido de Melo Carvalho e Dante Teixeira.
2. I use the phrases 'native American' and 'Indian' interchangeably here, since in Brazil the word 'Indio' is always used and the phrase 'americanos nativos' would refer to US citizens, not Indians.

3. For a recent bibliography on the Brazilian Indians see Seeger, 1980, Chapter 7; for an introduction to the Brazilian Indians in general, see Melatti, 1972; for a history of their contact with national society, see Ribeiro, 1970; for a discussion of the effects of development on some of the groups indicated see Davis, 1977; for up-to-date reports see Anthropology Resource Center, 1979 ss.
4. Most of my information for this section comes from the paper cited (Agostinho da Silva, 1978) and from personal communication with the author; another important source for further clarification of the Pataxó situation would be Carvalho (1977). José Candido de Melo Carvalho (personal communication) also clarified some important points.

REFERENCES

Agostinho da Silva, P. (1978). 'Fundamentos para o estabelecimento da reserva Pataxó'. Mimeographed Reproduction. Reprinted in *Revista de Antropologia* 23:19–29. São Paulo, 1980.

Anthropology Resource Center (1979 ss). *ARC Bulletin*—A bimonthly report on the Indian situation in South America. Boston: Mimeograph.

Carvalho, J. C. (1981). *A conservação de natureza e recursos naturais na Amazônia brasileira.* Companhia Vale do Rio Doce, *Revista*, Edição Especial, novembro de 1981.

Carvalho, M. R. (1977). *Os Pataxó de Barra Velha. Seu subsistema econômico.* Salvador: Dissertação de Mestrado de Pós-Graduação em Ciências Humanas da Universidade Federal da Bahia.

Davis, S. (1977). *Victims of the Miracle, Development and the Indians of Brazil.* Cambridge: Cambridge University Press.

Goldman, I. (1963). *The Cubeo.* Urbana: University of Illinois Press.

Meggers, B. (1971). *Amazonia: Man and Culture in a Counterfeit Paradise*. Chicago: Aldine.

Melatti, J. C. (1972). *Índios do Brasil.* 2nd edition. Brasília: Coordenada Editora de Brasília.

Presidência da República. *Polıtica Indigenista—Estrutura Jurıdica.* Série Documentos Sociais. Brasília: Gabinete Civil.

Ramos, A. and Taylor, K. (1979). *The Yanoama in Brazil 1979.* IWGIA Document 37, Copenhagen: International Work Group for Indigenous Affairs.

Reichel-Dolmatoff, G. (1976). 'Cosmology as Ecological Analysis: A View from the Rain Forest.' *Man*, volume 11, no. 3, pp. 307–318. London: Journal of the Royal Anthropological Institute.

Ribeiro, D. (1957). 'Culturas e Línguas Indígenas do Brasil'. *Educação e Ciências Sociais*, volume 2, number 6. Rio de Janeiro.

Ribeiro, D. (1970). *Os Índios e a Civilização.* Rio de Janeiro: Editora Civilização Brasileira.

Seeger, A. (1980). *Os Índios e Nós; Estudos sobre sociedades tribais brasileiras.* Rio de Janeiro: Editora Campus.

Seeger, A. (1981). *Nature and Society in Central Brazil.* Cambridge: Harvard University Press.

Simonian, L. T. (1981). *Terra de Posseiros; Um Estudo sobre sobre as políticas de Terras Indígenas*. M. A. Dissertation. Rio de Janeiro: Museu Nacional.

World Bank (1981). *Economic Development and Tribal Peoples: Human Ecologic Considerations.* Office of Environmental Affairs. Washington, D.C.

Wright, R. (1980). *History and Religion of the Bainwa Peoples of the Upper Rio Negro Valley.* Ph.D. Dissertation, Stanford University.

Socio-economic Effects and Constraints in Tropical Forest Management
Edited by E. G. Hallsworth

17

Current management concepts in forestry

D. N. MISRA
Indian Institute of Forest Management, Ahmedabad

INTRODUCTION

There was a time when forests were plentiful, human populations small and man's needs for timber, firewood and other forest produce modest. Then came industrial revolution, expanding human populations, increasing demand on timber for ships in the navy, and later for railways. Forests started being cut down for timber needs, for providing residential and industrial complexes, for iron-ore smelting and the like, and the fear of timber famine led to the need for protecting forests and regulating their fellings in such a way that forests would continue to yield timber and other forests' produce to sustain human needs. This marked the beginning of forest management. The initial steps, almost everywhere, consisted in regulatory statutes for protecting endangered forests and species. For example, there were regulations against forest fires, reserving certain species for specific needs, like white pine for ship masts, forbidding trees of certain size and species from being cut by private individuals and reserving them for public use, demarcating forests as statutory reserves to be controlled and 'managed' by the government, and so on. Scientific forest management, as we know today, was still some distance away, but it may be noted that its beginnings were centred in the concern for protection and conservation of trees and forests and this was to have a very profound influence on the course of evolving forest management.

Prohibitory regulations took different directions in different countries. In some countries, like India and Canada, most of the commercially important forests were taken over and put under state control; in some others, like the USA, Sweden and Japan, considerable areas of forests continue to be in private hands with various mechanisms evolved for some kind of overseeing of their management by the governments. Initially there was bound to be considerable divergence of views on the management objectives of privately owned and state-owned forests, but in later years these have, more or less, converged.

TRADITIONAL FOREST MANAGEMENT

Sustained yield

As mentioned earlier, the state was concerned in regulating the fellings in forests in such a way that forests would continue to yield timber and other forest produce in the future. This led to the concept of sustained yield, and it has since then remained one of the pillars of forest management. Initially 'sustained yield' was meant to convey undiminishing yield—that is managing forests in such a way that the existing level of outturn is at least maintained to meet the needs of the society. This was decidedly the first concern of forest management in an era when forests were threatened due to unplanned fellings, and shortages started occurring in some special areas like timber for shipbuilding. In India, too, the earliest plantations of teak in Nilambore were carried out in 1842 to meet the threatened shortage of teak timber for the British Navy. However, 'sustained yield principle' over the years became such a fetish with foresters that it became a kind of fetter that chained down foresters and forest management for well over 100 years and prevented them from assuming a more flexible and dynamic approach towards production. Even now, there is controversy as to what should replace the 'sustained yield principle'. The whole edifice of traditional forest management would collapse if the foundation of sustained yield were eroded or replaced. Lately, some countries have replaced the 'sustained yield principle' in their forest policy resolutions with the words 'progressively increasing yields' to counter the impression or tradition that sustained yield was meant to be a stationary figure of production, and if higher yields were taken out in certain years they had to be compensated by reductions of fellings in subsequent years or in other areas of management units in the same years.

Normal forest

The concept of a normal forest was corollary to the principle of sustained yield. Yield from an over-felled degraded forest could be very low or again, if most of the crop was young and not ripe for harvest, the yield would be nil. The 'sustained yield' principle would obviously not apply to such a forest. Foresters then invented the concept of a normal forest. Theoretically, it is a simple principle and wholly understandable; all age gradations from year one to the rotation age being present over equal areas. Each year one area would cross over to the rotation age and would be felled, giving an annual yield (which is the sustained yield). In practice, however, such age gradations would be possible only in a plantation forest. It is very difficult to ascertain the 'normality' or otherwise of a natural forest, and a mixed forest at that! The difficulty even in a plantation crop is to define and lay down the 'normal stocking' in each age gradation. Empirical standards were laid down by creating sample plots in as even-aged natural crops as possible and counting the number of stems growing in a healthy state (without being suppressed)—it being assumed that the area selected does not contain any gaps or blanks.

In spite of over 100 years of scientific forest management in India, where one of the objectives of management has, throughout, been the creation of a normal forest, there is hardly any example of a normal forest having been created anywhere. The concept has remained theoretical and largely impractical. However, the normal forest and sustained yield have been the twin pillars around which the edifice of traditional forest management was built; and in spite of being attacked and called into question lately, they continue to be the basis of forest management as before.

Silvicultural systems

Unlike the twin concepts of 'normal forest' and 'sustained yield', the third important ingredient of forest management, viz. the silvicultural systems, has a sound basis rooted in field experiments and refined continuously over the years. The aim of a silvicultural system is perpetuation of forests through constant renewal of crops, and therefore they revolve around the methods of regeneration of forest crops. There are two basic ways of doing so; (1) leaving and helping nature to regenerate—called natural regeneration, and (2) through plantings—called artificial regeneration. Classical forestry has a large number of well-defined silvicultural systems evolved by German and French foresters based on various ways of inducing natural regeneration—most of them irrelevant to forest crops in India. The methods in universal use are the Selection System and the Uniform Shelterwood System, helped by artificial regeneration wherever necessary.

Yield calculation and regulation

Yield calculation and regulation is the fourth important component of forest management. In a normal forest, the yield calculation is quite simple—the mature crop at the rotation age is felled and constitutes the yield. Calculating the yield from a natural crop is a complex affair and various methods have been devised for calculating the yield based on enumeration of growing stock in various diameter classes, rate of mortality, number of years taken by the crop to pass from one diameter class to the next, and the number of mature and overmature stems. Once the yield is calculated by any of the methods or combination of methods, yield is regulated by either volume or area or both.

Rotation

One of the crucial decisions to be taken in forest management is the fixing of a rotation age. There are many considerations in fixing a rotation, the most important being the end-use to which the crop is to be put—i.e. whether the object is to grow saw timber, poles or pulpwood. The rotation age is determined from the growth curves available for various species. Where, however, all dimensions of timber have a ready market, rotations have to take into account the price differentials of various diameter classes and the volume production at various rotations. Three kinds of basic rotations are commonly visualized—the physical rotation, the financial rotation and the rotation of maximum volume production. In arriving at an actual decision about rotation, however, a number of factors have to be integrated like (1) financial objectives, (2) forest factors like silvicultural, entomological, pathological, and site quality and (3) the product values in the market.

There are, of course, other fields with which forest management has been concerned, like wildlife, recreation, villagers' rights and concessions (such as grazing, fuelwood, hutting and agricultural timber), fire control, and the like. But these have been, until recently, merely peripheral activities, whilst the core of forest management has remained silvicultural systems, yield, rotation, growing stock, stand structure, growth data of species and nursery techniques.

MODERN DEFINITION OF FOREST MANAGEMENT

Roth (1925) defined forest management as the task of 'build[ing] up, put[ting] in order, and keep[ing] in order a forest business'. The Society of American Foresters (1958) has elaborated the same, and defined forest management as 'The application of business methods and technical forestry principles to the operation of a forest property'.

Davis (1966) has listed the following subjects with which forest management is concerned (see Table 1). This is a pretty exhaustive list of the wide areas with which forest management is intimately concerned. While administration, finance, accounting and statistics have been part of forest management in India and foresters have been trained in these subjects (except for finance which was considered only part of accounting) the other subjects listed under business and social aspects have hardly ever been included as part of forest management. Economics got into the curriculum of forest colleges in the decade beginning with 1970 and the business and social aspects came into focus with the establishment of the state forest corporations around 1974–75. With the dwindling of fossil-fuel oil and the rise in its price, the decade of the

Table 1

Business and social aspects	Technological aspects
Economics	Silvics and silviculture
Organisation and administration	Mensuration (and statistics)
Finance	Logging and saw milling
Accounting	Wood technology
Statistics	Pathology
Marketing	Entomology
Business law	Fire control
Labour relations	Wildlife and recreation
Real estate	Grazing
Social and political science	Water
	Civil engineering

1980s will see forests as an important renewable energy resource and the best utilizer and harnesser of solar energy. Along with this is the current concern of humanity for environmental degradation and pollution, and the continuous and frightening rise in populations of the developing countries. This adds new dimensions to the priorities in forest management.

INTERFACE BETWEEN FOREST MANAGEMENT AND FOREST POLICY

Apart from the technological aspects of management of forests which are independent of the objectives of management (as they are concerned with the biological and physical aspects of growth of trees and crops), the latter play the all-important role in determining how the forests are to be managed. And the objectives of management will depend on the forest policies of the state. The first forest policy statement was issued in India in the year 1894. At that time forests were still plentiful, and the basic approach then was that agriculture had priority over forestry and whenever and wherever land was needed for agriculture it would be surrendered without protest. The other overriding principle was that forests were to be managed primarily for the benefit of the villagers whose needs would have priority over other uses. Of course, the forests were to be managed on the sustained-yield principle, and efforts were to be made to see that the villagers were taught the need for proper forest management and were dissuaded from over-using and degrading them.

REVISED POLICY OF 1952

The 1894 policy was revised in 1952. During the intervening 58 years the forest picture had changed dramatically. From an abundance of forest resources the net area under forests had shrunk tremendously, forests had in general been over-felled during the Second World War, and had suffered from the after-effects of independence and partition. There was need to revise the policy priorities. Forestry had to be put on a separate pedestal and recognized as having a legitimate claim to land and not merely allowed on sufferance to agriculture. A large number of industries based on forest raw material had come up, and these were to be fed with forest produce. The principle of a fixed sustained yield had to give way to the dynamic concept of progressively increasing yields until the full productivity potential of the site was reached. Private forests had to be controlled so that their owners did not liquidate them and turn over the land to the land-hungry agriculturalists from across the partitioned borders. Forests were to be looked upon as multiple-use resources and in order to have proper focus on management with respect to priorities in the multiple-use concept, they were to be classified as commercial, protected and village forests. Environment had yet to make its impact on the world consciousness.

The revised policy of 1952 made its impact on the regeneration methods of forest management. Emphasis had shifted to artificial regeneration in areas where natural regeneration was slow, uncertain, or not possible for a variety of reasons; biotic, soil and others. Also species that had remained more or less of secondary importance were in demand and plantations on a large scale started from the middle 50s. Blanks, degraded forests, eroded areas, and non-regenerating forests were all put under plantation crops; and along with them came the practice of what is variously termed as agri-silvics, taungya, or forest farming. Agricultural food crops were raised in the intermediate lines which gave a boost to food production and helped in the maintenance of plantations in the early stages.

Another important policy statement contained in the 1952 forest policy was that it should be the aim of the states to maintain at least 33 per cent of their area under forests and wherever it was deficient, steps were to be initiated to make up the balance. This also helped to develop plantation activity and in resisting the increasing pressure from various quarters for surrendering forest lands to agriculture.

To meet a vastly changed forestry environment of today, the forest policy revision is again under query. It is bound to reflect the current concerns of environmental conservation, energy crisis, pollution, and the role of forestry in integrated rural development. Social forestry seems to have become the catchword of the present decade and a panacea for the many ills of the village and rural poor.

FOREST MANAGEMENT AND INDUSTRIAL NEEDS

Nowhere is the area of conflict deeper than on the issue of policies for forest development and the use or abuse of forests for industrial needs. The decade beginning with 1950 in India was one of rapid industrial development, and this tempo has been, more or less, maintained since then except for shifting emphasis on heavy, medium- and small-scale industries. A consequence of this has been a steep rise in the demands for products based on forest raw material—plywood and veneer; pulp and paper; packaging; saw timber; and writing, printing and all other kinds of paper, including newsprint. The National Commission on Agriculture, in its comprehensive report on forestry, brought out statistics which showed a massive gap between supply and demand in the future years. It laid emphasis on a changeover from conservation-oriented to production-oriented forestry and a massive investment to raise productivity of the forests which was amongst the lowest in the world. State forest corporations were formed in major states to take advantage of financial institutions for the massive forest plantation programmes which aimed at replacing long-rotation, low-yielding natural forests with short-rotation, high-yielding tree crops (man-made forests). In short, a strategy for forest development was adopted which was more or less similar to the one that brought in the green revolution in the agricultural field. But forestry is not agriculture. Natural forests with mixed species, and with a stable ecosystem, have their own value and the conservationists, ecologists and environmentalists are all up in arms against the 'dangerous' and 'short-sighted' policy of replacing natural crops wtih either exotics or even with indigenous monocultures. The economics of such a step is itself being challenged, for it is argued that the maximum biomass production on a certain site in perpetuity is possible only in a stable ecological climax community. With the dangerous levels which pollution has reached in the urban areas, people everywhere are against any step that would lessen the forest values—as they see them—wilderness, wildlife, oxygen bank, soil and water conservation, atmospheric filter, climate amelioration and so on. The argument is that technology has to change and adapt itself so that the entire biomass produced in the forest areas naturally can be more efficiently utilized than trying to grow only a particular species for the utilization of which technology already exists. The arguments on both sides are weighty and the controversy goes on. Sometimes it takes on the form of 'Chipko Movement' which was launched in Uttar Pradesh to save the hill forests from being felled to feed the hungry papermills. And this brings us to another important dimension of forest management.

FOREST MANAGEMENT AND PEOPLE

Ultimately, the aim of forest management is to serve society—to supply goods and services which are somewhat unique in nature and can only be produced through proper maintenance and development of forest resources. True, in certain fields like artificial fibres and packaging materials, plastics have taken over, but in the rest of the vast field of forest products, wood and the living forest are still the answer. When forest management in India started, people's needs were the focus and the first priority. However, as forest industries grew and the demand for forest produce increased, the focus shifted to industries and what generally came to be known as commercial forestry. Occasionally there were agitations from people, especially in the Himalayan region where forests are intimately woven with the everyday life of the villager. To pacify their demands, large chunks of reserved forests were taken out of the control of forest departments and entrusted to the district collector (Uttar Pradesh in 1921). As forests became scarce due to continued deforestations (especially after independence), and the demands soared—both of industries on the one hand and of increasing village populations on the other—severe competition set in between the industry's needs and the villager's demand. The forest departments came under severe criticism as people thought that the forester's first concern appeared to be the forest industries and revenue from sales and not the villager's needs and demands. This criticism of the villager is reinforced by various other groups which have come to take a keen interest in the proper management and maintenance of forests. Such groups are the wildlife enthusiasts and scientists, environmentalists, tribal and rural welfare and development organizations, small-scale forest-based industries, groups and organizations interested in renewable sources of energy, anti-pollution watchdog associations, and groups interested in urban forestry. The criticism of the forester has reached such proportions and his condemnation is so universal, that it would appear that all his valiant attempts at forest conservation and his highly developed forest management technology seem to be suddenly inadequate. Whether it is the forest industrialist, the elected representative of people in legislatures, the tribal, the villager, other government agencies and departments concerned with forests like animal husbandry, agriculture, irrigation and power or the newly formed groups mentioned above, none has a good word to say about foresters and their attitudes. Each, of course, has its own reasons which are often contradictory to each other.

There is certainly a large amount of truth in the allegation that the forester's emphasis has shifted and indeed, long been shifting from the villager to the industries. But in this it is not he who is to blame. The National Commission on Agriculture roundly condemned the forester's conservation-oriented forest management and castigated him for keeping forest productivity amongst the lowest in the world. It recommended intensive forest management and strategies that would push up forest productivity in the short and the long run. Some of these strategies run counter to the demands on forests of others, and the forester is caught up in the crossfire of the opposing parties. He has sympathies of neither and is thoroughly confused—often paralysed. Here is a case for a need to lay down a clear-cut forest policy so that the forest management could be geared to achieve the objectives of that policy and there would be no groping in the dark for the practising forester.

FORESTRY AS A BUSINESS

Although foresters in India have always emphasized forestry as quasi-commercial in character, it never was, and is not now, thought of as a 'business'. The forester has regarded himself a technologist and an administrator—seldom a resource manager; and whether or not forestry is regarded as a 'business', the business of forestry certainly is concerned with 'management'—management of a resource. Technology is certainly the foundation of forest management, but that alone is not enough. A whole lot of other sciences, disciplines and bodies of knowledge are an essential part of the job of management. Davis (1966) has divided forest management into two parts: technology, and business and social aspects. Forest management could perhaps be more succinctly defined to include (1) technological aspects; (2) general management. The word 'general management' would include all those aspects which Davis has listed under business and social categories. The forester has to be a good technologist and also a good manager. In fact, part of the 'failure' of the forestry system and foresters as perceived by its critics is not of technology, but his managerial side. For long the forester has been devoted to his forests and forestry in complete isolation of the happenings in the external environment. He failed to take a note of these—in fact, almost disdainfully ignored them—till nemesis overtook him. It is only from the beginning of the 70s that the forester has grudgingly started recognizing that there is strong action and interaction between forestry and its vast external environment, and he could no longer keep forestry from being polluted by such social sciences as economics, social psychology, sociology, political science, organizational behaviour and the like; that the art and science of forest management in its traditional connotation is affected and has to be continually adjusted to the changing needs, moods and beliefs of the society. There is strong correlation between forestry and society—what happens to society affects forestry and the changes in forest management affect society. It is for this reason that the forester has to have a system of continual monitoring of social changes, an insight into social behaviour and social psychology and the current political changes. For the forester the 'good old days' of sheltered unconcern are gone for good and his job has become ever more complex. Without the insights provided to him through general management education he cannot hope to succeed. The definition of forest management has to include technical forestry and general management principles.

CONCLUSIONS

What does all this lead to? The first is a recognition that the traditional and classical concepts of normal forest, sustained yield, silvicultural systems, etc. on which forest management has so long been based will not do. The basic objective of managing a forest resource is to meet societies' needs—be it production of raw materials for individual and commercial use and/or other goods and services. The yield has to be a function of society's preferences and the forests have to be managed to meet those needs and not for the aim of achieving a 'normal forest'. There is no sanctity of 'sustained' yield. The yield can and would vary on the exigencies of the situation. At any given time, the growing stock would certainly be capable of giving a certain 'maximum possible' yield. It is certainly a desirable goal that society should not suffer privation in the future by dwindling yields. For this reason the entire crop structure and growing stock has to be analysed periodically, management models constructed and tested to forecast the future behaviour of yields and the concerned parameters manipulated to allow the yields to move in the desired directions.

There has to be a greater precision even in the biological sphere of forestry. The large number of silvicultural systems taught at the forest colleges and schools have become largely irrelevant. All management systems which have a large element of inbuilt uncertainty and unpredictability have to be scrapped. The age of the long-drawn-out regeneration periods of the crops is over. If natural regeneration cannot be assured within a definite time frame, artificial regeneration has to be adopted.

Forest management is not merely forestry—that is managing and growing forests. Although the beginnings of forest management were rooted in the conservation of forests, and until a few decades ago, this was all that the forester was concerned with; over time, the term has come to mean much more. Apart from the physical and biological aspects the social sciences have a very intimate relationship with forest management. While none of the basic forestry data such as growth rates of trees, physical rotation, pests and diseases, and silviculture of trees is affected by the external environment, the management strategy is largely dependent upon the overall national policies; and rotation, choice of species and silvicultural systems may, in most cases, be the best compromise between forestry and the social needs and demands. Management education that deals with social sciences has, therefore, to be as much a part of the foresters' training as technical forestry.

The old forester-syndrome in the management of forests—the pride in doing everything by oneself and the image of an all-knowing forester—has to be discarded. As shown in the preceding sections of this chapter, forest management affects and is affected by a large number of factors in the external environment. To follow and to practise an effective and relevant forest management regime, a clear appreciation and understanding of the forces operating outside of forestry spheres is essential. It is beyond the capacity of even the normally intelligent forester to grasp and master all the social sciences with which forest management is intimately involved. Help from experts in these fields is, therefore, essential in deciding upon and evolving the best strategy of forest management. Foresters should not fight shy of creating a management pool of such experts within their own planning and managerial set-ups to advise and guide them. This alone can ensure the relevance of foresters and forest management to the society and the country, and the continual adjustment of strategy to the changes occurring in the external environment.

REFERENCES

Davis, K. P. (1966). *Forest Management: Regulation and Valuation*, 2nd edn.
Government of India (1976). *National Commission on Agriculture*—Part IX.
Roth, F. (1925). *Forest Regulation*, 2nd edn.
Society of American Foresters (1958). *Forestry Terminology*, 3rd edn.

Socio-economic Effects and Constraints in Tropical Forest Management
Edited by E. G. Hallsworth

18

Forestry as a social institution

T. BURGESS
North-East London Polytechnic, Romford Rd, London, UK

One of the components of the 'Save Our Soils' project of the International Federation of Institutes of Advanced Study is a series of desk studies of the land tenure, soil conservation and forestry legislation in the participating countries. The desk studies will assemble sources of material available in England and from the collaborating countries, analyse appropriate legislation reflecting a variety of problems and practice, prepare comparisons of types of land and forestry law and discuss the effect of such law on the management and use of land and forests. This work is being undertaken at North-East London Polytechnic, and this chapter sets out the principles by which the problem is being studied.

The first principle of our approach is to regard the law, and indeed customary practices, in the same light as other public institutions; that is as an implicit or explicit solution to some problem or problems. This immediately raises the question to what problem or problems a particular piece of legislation or public institution is a solution. It has the merit of directing attention to questions external to the institution itself and makes possible the development of criteria against which it can be judged. The approach has already proved fruitful in a number of fields, including local government, public finance, education and health. It has been applied to some aspects of the law in England. Before considering forest law and forestry institutions, therefore, it is important to set out the major problems in forestry, together with existing or proposed solutions.

Perhaps the most familiar problem is that of governments: how to secure consistent supplies of timber, to serve the interlocking priorities of defence and communications. The solution to this problem has been sought in protective legislation. Areas of forest have been reserved for production, felling in them has been reserved to the government and the areas themselves are policed by a forestry service to prevent depredation. Where necessary these reserves or production forests have been supplemented by plantations. A second problem, that of the protection of watersheds and other areas against erosion, has been increasingly recognized, and although very different from the first has been met with similar solutions. Protection forests have been established or designated, and governments have extended to them 'reserve' measures and policing.

In recent years a general desire to protect forests has arisen even when the consequences of their destruction are not as immediately obvious as that of watershed forests. There is a growing argument that the consequences of widespread clearing, though often unpredictable, may be deleterious. It is increasingly asserted that the preservation of forests is a general problem. Unfortunately, the destroyers of forests are very diverse. Governments themselves may be involved through a general pressure for 'development' which would replace forests with agriculture, industry or 'infrastructure' schemes, for example for hydro-electric power.

Another source of pressure is the market. The demand for timber leads to widespread felling. This can be organized on a large scale, by national and multinational corporations, and may thus be hard even for governments to resist. Yet another pressure may arise from the growth of population in or near forest areas, leading to destruction by clearing or overuse. Solutions to those problems are as varied as the problems themselves.

Governments seek to balance competing demands within their own structures by means of co-ordinating agencies, public enquiries and the balancing of pressure groups. They try to control large corporations by systems of licensing. They react to population growth with 'reserve' measures, at one extreme, to deportation at the other.

The problem of preserving forests develops in some countries into a problem of extending them. Governments may wish to reclaim land which has fallen into underuse, disuse, or even desert. Solutions to this problem vary from large-scale planting to proposals for the encouragement of trees by local people on quite small plots of land—a solution which is coming to be known as 'kitchen forestry'. What all these problems have in common is that they are formulated by governments or their agencies, and solutions for them tend to be sought on a national scale.

There are a number of other problems, which are not so much those of governments or countries as those of people who live in or near forests. The problems here may be quite basic: those of maintaining a bare livelihood. Some of the solutions of the traditional forest-dwellers, like shifting cultivation, have proved acceptable where the pressure of population is light but have increasingly been seen to be self-destructive. In such places a solution has been sought in such notions as social forestry, where forestry is combined with agriculture or grazing and where the attempt is made to persuade people to avoid the successive destruction of parts of the forest as a way of life. There are also the problems of those who live near rather than in forests. They may be villagers in a predominantly agricultural society, but they use the forest for such necessities as building material, fuel and food for people and animals. Their problems are exacerbated both by the destruction of forests and their over-rigorous preservation. There have been a number of spontaneous movements by local people, of which the 'chipko' or 'tree-hugging' movement in Northern India is the best known, which have taken direct action against local tree-felling. Such problems suggest solutions in terms of right of use of forest products. In a number of countries such rights, particularly for tribal or backward peoples, are enshrined in constitutional guarantees.

There is also a growing belief in the use of forests for recreation and for the proper balance of the national habitat. Solutions to the problem of maintaining forests for these purposes are sought in the development of national parks and protected areas.

Finally there are those who argue that the preservation of forests is a global problem not just in the sense of universality but because the disappearance of forests may cause serious climatic and atmospheric changes. Such people look forward to the development of international institutions as a solution to a world problem.

In the present debate about these problems and their solution it is possible to identify a number of positions taken up by participants of different kinds. There are what may be called the traditional foresters, with well-established practices for policing and preserving reserve or protection forests. There are a growing number of academics and bureaucrats with a professional interest in the problems of the people who live in or near forests. There are those who see the forests as part of a 'total ecosystem'. These groups are not necessarily mutually exclusive. There are many traditional foresters who seek to accommodate the forest-dwellers. Academic sociologists and ecologists may recognize the need of national governments for secure supplies. Economists take account of politics. But it is probably true to say that in most countries the conflicts of interest in the control and use of forests are as yet imperfectly resolved.

The developed forestry service of the future, in any country, will have to offer solutions to the problems outlined above. The problems differ. The solutions may be incompatible. Interests conflict. A central, additional, problem for a forestry service, therefore, is to make possible the resolution of conflict, to accommodate within itself the means of this resolution. Much forestry legislation in the past, and many forestry services, have offered solutions to a relatively narrow range of problems. Both may have to develop so as to accommodate not only a wide variety of different kinds of problem but a resolution of the conflict between different kinds of solution. Fortunately, in the development of public institutions, there is considerable experience of creating both law and structures, and it is worth learning from this experience in the service of forestry. Public services like health or education have to offer solutions to the problems of individual users—patients or students and their families; of the professionals who offer the service—doctors and nurses or teachers; of those who administer its provision; and those of the people at large as taxpayers or electors. Some of the solutions to these problems will clearly be in conflict. There are even conflicts of interests within the various groups, between the needs of different kinds of patients, for example. Such conflicts cannot be settled once for all. An education service or a health service has to contain within itself the means for their resolution at different times and in different circumstances. Similarly a developed forestry service must accommodate the continuous resolution of conflict between users, providers, professionals and the nation at large.

A number of countries are already evolving such means of resolution. Traditional forestry services are changing. Regional and local governments are being given new responsibilities. Small communities and villages are being invited to make decisions, within an overall framework, where in the past they would have been required simply to comply. Efforts are being made to encourage responsibility and participation. Most of the development is proceeding, as it must, by trial and error. It may make the trial more fruitful and the error less damaging, however, if there is some broad theory upon which new laws and institutions can be built.

Perhaps the major function of new forestry legislation should be to provide for the apt distribution of powers and duties. Different functions can be performed at different levels of administration or by different bodies. Similarly, accountability can be secured at every level and institutions organized so as to encourage democratic participation and control. These major questions may be decided differently in different countries or circumstances but the issues themselves do not change.

We begin with the ministry responsible for forests. It is important that legislation should not overburden a minister. His powers and duties should be those which can rationally be undertaken by an individual on the advice of his department and for which he can reasonably be held accountable to the Parliament or other legislature. A minister needs to be given the general duty to secure the development of forestry and to evolve the national policy for it. He needs sufficient reserve power to maintain a general, but not detailed or day-to-day, control over responsible bodies and he needs the power of inspection. He will inevitably be engaged in determining the overall resources available to forestry as a result of his battles with ministerial colleagues.

One of the most important policy decisions for any minister must be the extent to which the development of forestry is to depend upon the market on the one hand and the workings of democratic politics on the other. This is not something that can be decided on first principles, or once and for all, but must be a matter for each minister, according to his judgement of the circumstances. The minister's legal powers and duties should not be couched in such a way as to preclude or prejudge decisions of this kind.

The next tier below the minister will be various kinds of statutory bodies. It is an important matter of decision here how such bodies should be established. In some countries the forestry service is simply a part of a central ministry, under the direct control of the minister. The disadvantage of this arrangement is that the minister and ministry can be subject to little external check, and it is hard to secure accountability for detail through a central legislature. It also makes it impossible for a minister to adjudicate in disputes or act as a court of appeal.

In other countries forestry is managed by a statutory forestry service or forestry commission which has statutory responsibilities separate from those of the minister or ministry. The relationships between such commissions and the government depend upon the degree of independence they are given by the instrument which establishes them and the extent of any reserve powers vested in the minister.

The history of these statutory services carries mixed lessons for the future. On the one hand they can be said to have been extremely successful in the narrow traditional task of forest policing and preservation. They have been less successful in gaining acceptability among local people. For this latter reason the question arises whether it is possible to introduce any popular influence at a local level of operation. One institutional device for doing this would be to establish committees charged with the oversight of local decisions of a statutory service. Such committees might have a substantial representation of local people either directly or through their local representative institutions, the staff of the service and the central commission or board. The committees could be entrusted with a range of powers and duties including local policy (within overall guidelines) and finance (within an approved budget). Their powers and duties would be incorporated in articles of authority which would make clear their relationship with the central body and would make it possible for the local or divisional organization of the service to be accountable, without confusion both locally and nationally.

A third kind of statutory body might be the normal organs of provincial, regional or local government. This use of local institutions, under the general direction of the minister, might be an alternative to the establishment of separate statutory services. The most important danger to be avoided here is in creating several tiers of forest administration corresponding to different levels of regional and local government. The most recent developments in forestry suggest the creative use of such institutions, as a way of encouraging and legalizing social forestry or the rational incorporation of rights to use (for fuel, food, fodder and so on), in traditional forest management. There are also those who believe that a more local democratic influence in forestry would counter a pervasive and destructive urban dominance of rural life.

A new forest law, in short, could establish local authorities as responsible or providing bodies for forests, responsible in general terms to the minister, but directly accountable in the normal way to a local electorate. By these means even substantial decisions, like the granting of licences for felling to large firms or the establishment of country parks, could be made amenable to the expression of local feeling.

Such are the institutional questions that need to be discussed in any attempt to create or administer forestry legislation. The temptation, when legislating, is always to seek to enshrine in the law particular policies. This is usually a mistake. It makes necessary changes in policy too slow and complicated. What is needed instead is a series of institutions whose different powers and duties enable different decisions to be taken at appropriate levels. For example, existing forest legislation in many countries is biased towards the policy of preservation. It impedes the expanded use of forests, the growth of local responsibility and the encouragement of activities (like agroforestry) which were once regarded as undesirable but are now widely recommended. In all these ways the law sustains particular policies. It is important not to make the mistake of amending the law simply to sustain once-new policies even when better ones present themselves. In other words, the problem of forestry law is not to enforce ideas but to create a new social institution.

Socio-economic Effects and Constraints in Tropical Forest Management
Edited by E. G. Hallsworth

19

Report on FAO activities

J. E. M. ARNOLD
Forestry Department, Food and Agriculture Organization of the United Nations

INTRODUCTION

In a paper, entitled 'Towards a forestry strategy for development', which was put before its Committee on Forestry in May 1980 (FAO 1980), FAO has argued that a basic requisite for dealing with the problems of excessive deforestation in the tropics and elsewhere in the developing world must be forest policies which add a social objective to the usual productive and protection objectives.

Forest development strategies that stress only the production of industrial wood, or only the productive and protective functions of forestry, are inadequate in a number of ways. Firstly they usually ignore or give low priority to the needs of people living in or adjacent to the forests—which are usually for outputs other than timber—such as fuelwood, grazing, bush meat, etc. This dichotomy of interest, when accompanied by measures to restrict local access to or use of forests, goes far to explaining the negative behaviour and attitude of people towards forests in many places. They do not perceive the forests as being something which exist, and hence should be conserved, for their own benefit.

Policies which focus only on the timber production and protective functions of the forest are also inadequate in a situation of growing demographic pressure because they exploit only part of the potential of forests and forest lands to produce needed goods and services. Timber crops are widely perceived, often even by national and agriculture sector planners, as of less value than the agricultural crops that could be grown on the site. The protective values of the forest cover are either misunderstood or ignored—overridden by what are seen, rightly or wrongly, as the higher values to which the land can be put under food or cash crops, even where this is a short-lived and destructive use of the soil on the site. The only hope of averting such deforestation in areas where forest is the best, and often only, possible long-term land use would seem to be by exploiting the potential of trees to produce more than timber, of forests to be a habitat for more than just trees and of forest lands to support the production of more than forests.

A major part of FAO's effort in the field of forestry at the present time is hence to bring about awareness of the need for broader, more development-oriented, strategies and policies in forestry, and commitment by governments to such policies, and to assist them in implementing them. The new directions essentially seek a holistic approach to the management of forest lands, in which their potential to contribute to local needs for food, fuel, income, etc., is integrated with their timber and environmental outputs. Central to its success is a stronger commitment by governments to forestry as a tool for rural development, and a fuller involvement of rural people in the management of, and benefits from, forests.

This approach to forestry is the subject not only of the FAO paper cited above but also of a special chapter entitled 'Forestry and rural development' in FAO's annual report, *The State of Food and Agriculture*, 1979. The latter is illustrated with accounts of a number of innovative attempts to apply such an approach which have already produced encouraging results, or which show signs of doing so.

INFORMATION FOR FOREST LAND-USE PLANNING

A major impediment in dealing with forest management in the tropics has been the paucity of reliable information about either the extent or nature of the forest, or about its rate of depletion. The recent UNEP/FAO Tropical Forest Resource Assessment Project was designed to provide such information at regional and global levels, and to define a methodology and means for the continuous updating of this first assessment.

The results of the project (FAO, 1982) indicate that the area of closed tropical forests is diminishing by 7.47 million ha annually—4.34 million ha in tropical America, 1.83 million ha in tropical Asia and 1.33 million ha in tropical Africa. Open tropical forests are also decreasing in area, by 3.81 million ha annually. Shifting agriculture is by far the most important cause of this deforestation, accounting for about 46% of the total.

FAO's experience in land classification for forestry ranges from such global level studies to the project level. At the latter level work has tended to be designed in response to locally defined needs, and standardized land-evaluation principles and methods have not been widely used. A need for a 'guidelines' document for land classification for forestry is recognized, and the possibility of preparing it within FAO's future programme is under discussion (Botero, 1981).

FAO has also accumulated extensive experience in forest resource surveys.

Recent and current work covers several facets of this (Singh and Lanly, 1979). A major focus is on adapting survey methodology to encompass the wider range of data needed for including fuelwood and non-wood products in forest management planning.

For the 1981 UN Conference on New and Renewable Sources of Energy, FAO assessed for each developing country—and within large countries for different parts of a country—the balance between fuelwood supply and demand (FAO, 1981). This study showed that about 2000 million people were dependent on fuelwood for cooking in 1980, and that more than 1000 million of them were living in situations where supplies were being maintained only by overcutting the wood resource. It also indicated that, if present trends continue unchanged, the number of persons in such situations will more than double by the year 2000.

TECHNICAL STUDIES

Central to the problems of forest management in the tropics is, as is well known, the lack of knowledge about how to manipulate their usually highly complex, and often fragile, forest ecosystems in a controlled fashion, or about the consequences of so doing. FAO has therefore been seeking to attract financing for programmes which would strengthen existing research entities in the different tropical forest regions and provide a network for intercommunication among them. Programme proposals of this nature have already been drawn up for the Latin American and the Asia–Pacific regions, but have not yet been funded.

Much attention is being paid in these and other aspects of FAO work to various ways of combining the growing of forest trees with agriculture. This is seen as being important not only in order to introduce stable multiple-use practices within the forest which increase the productivity of forest land; but also in order to extend the growing of forest trees outside the closed forests in order to reduce the pressure on the latter.

Growing attention is consequently being directed towards all aspects of tree-growing in predominantly agricultural areas. In the technical area this is focused on choice of species which best meet the constraints of land competition and needs for quick returns, and outputs of economic value to the grower, which are amongst the main impediments to tree-growing in these areas.

Current work in FAO in the field of agrisilviculture is concentrated on the assembly and dissemination of existing information. Review studies of agroforestry systems in South Asia and Central Africa are nearing completion. A compendium of information on important fruit-bearing forest trees is also in progress.

Multi-purpose forest tree species are also the subject of an FAO/IBPGR ongoing project on genetic resources of arboreal species for rural development. This concentrates on provision of genetic material of tree species which will be of value for planting by rural communities; and gives emphasis to arid and semi-arid areas and to production of fuelwood (FAO/IBPGR, 1980).

An overlapping area of work is that concerned with conservation methods and practices. The technical aspects of this have been the subject of the first five volumes of the FAO Conservation Guide series (FAO, 1976). The emphasis of present work is on the socio-economic and institutional measures needed for the viable introduction and application of conservation measures, which are discussed further later in this chapter.

Technical work in progress or planned on forestry activities which can be developed by local people as a source of income includes compilation and dissemination of information on production, processing and marketing of naval stores; medicinal plants; and utilization of wildlife for protein. A series of manuals on improved charcoal production technologies is also at an early stage of preparation.

SOCIO-ECONOMIC STUDIES

The upwards of 240 million people who are believed to live within tropical forests, and practise non-continuous cultivation, are amongst the people about which least is known. They largely escape censuses of population, and other demographic enquiries. FAO and the UN Fund for Population Activities (UNFPA) have, therefore, recently completed a first-phase study in a number of Asian countries (Bangladesh, India, Indonesia, Philippines, Thailand). The study, which was carried out in each country by a consortium of local institutions, was designed to develop more information about such populations, and about the factors that would affect the introduction of more productive and environmentally sound land-use practices.

Second-phase projects are now being prepared in order to introduce developmental activities in the situations studied. In addition, a similar study for Africa is under discussion with the UNFPA, together with a project to improve the relevant statistical base about forest populations on a global basis.

Another area of work is concerned with the gathering of information needed in order to plan forestry activities in such a way as to meet the needs of local populations. One component of this is a study under preparation, to appear under the title 'Rural development within the framework of watershed management' in the FAO Conservation Guide series, reviewing the range of information needed, and methods of collection and analysis, for upland conservation projects.

Another study in preparation, in the FAO Forestry for Local Community Development series, examines in more detail the process of identifying, information-gathering and analysing the often complex set of factors affecting the supply and use of fuelwood in situations where this is still largely a subsistence good. This is seen as particularly important where increased tree-growing is needed in predominantly agricultural areas, in order to accurately identify the constraints and possibilities for beneficial interventions.

Also in the general area of improving understanding of how to plan forestry activities in order to improve their positive impacts, and reduce negative impacts, on local people, is a study to define a methodology for evaluating the impacts of large forest industry complexes established in rural areas. It has long been postulated that these should have a strong local developmental effect, but there seems to be very little information about what actually occurs in practice. Should the methodology definition phase provide a usable output, it is hoped to commission a number of studies of actual recent forest industry developments in order to improve information in this area.

A further area of work concerns the provision of incentives, and removal of disincentives, to enable people to adopt sound forestry practices. Current work concentrates on incentives for community involvement in forestry and conservation programmes in upland areas, and will be published in the FAO Conservation Guide series with the title 'Incentives for people's involvement in upland conservation'.

INSTITUTIONAL STUDIES

An ongoing long-term comparative study of public administrations in forestry is being reoriented to focus on the changes that are likely to be required in order to administer and service forestry programmes with a strong rural development focus. A component study (Palin, 1980) has already identified some of the changes that are likely to be needed as forest services add to their traditional task of managing forests on public land a growing role of servicing others in growing trees on other lands. Two consultations on this subject, one in English and one in French, are planned for senior forestry personnel during 1982/83.

One element that has already been identified as needing widespread strengthening is forestry extension, to enable more effective support to be given to rural people in forestry activities. An English-language manual on forestry extension is under preparation. This will, *inter alia*, be used for a programme of training in this subject field, which started in January 1982 with a Workshop on the subject for countries in the Asia–Pacific region, in Indonesia.

Another institutional aspect that is under study is that of forest legislation. Most forest law still reflects earlier policies directed primarily towards conservation. It is therefore more a tool for preventing the misuse of the forest than a positive agent for development. A study is consequently under way to review existing forest legislation to determine in what ways it might need to be modified in order to remove constraints to, and encourage a broader-based approach to, forestry. The review is confined to forest legislation, though it is recognized that laws affecting land-use more widely may also require overhaul if the desired level of growing and use of trees is to be made possible.

The key to successfully adapting forest management so that it best meets socio-economic needs, and avoids socio-economic constraints, is evidently the proper integration of the technical, economic, social and institutional components of any situation. FAO is principally involved in this integrating process through the field projects it runs jointly with governments and funding agencies. For example, upland conservation projects in forest areas of considerable population pressure in northern Thailand, the Philippines, central Nepal, Haiti and north-east Honduras, or arid-land projects in several of the countries of the Sahelian zone, or projects to make management more effective, and relevant, in various countries within the high forest areas of all three tropical regions.

The experience gained through its involvement in such projects forms an important input into the FAO studies outlined earlier. In addition, FAO is undertaking a number of case studies of programmes which have successfully integrated socio-economic and forest management objective—the village forestry programmes in the Republic of Korea, and social forestry programmes of the state of Gujarat in India, and smallholder tree-farming programmes in the Philippines. It is hoped that these will provide more detailed information on, and understanding about, the requisites for success, and enable further elaboration both of some of the component studies, and of the overall process of managing forests in ways which will improve their contribution to rural development and to meeting the needs of rural people.

REFERENCES

Botero, L. S. (1981). *FAO's Experience in Land Classification for Forestry*. Proc. ISSS/IUFRO Workshop on Land Evaluation for Forestry, Wageningen, Netherlands.

FAO (1976 *et seq*)

FAO Conservation Guide No. 1—'Guidelines for watershed management', 1977.

FAO Conservation Guide No. 2—'Hydrological techniques for upstream conservation', 1976.

FAO Conservation Guide No. 3—'Conservation in arid and semi-arid zones', 1976.

FAO Conservation Guide No. 4—'Special readings in conservation', 1978.

Cahier FAO: Conservation des Sols No. 5—'Manuel de Contrôle des Avalanches', 1978.

FAO (1978). 'Forestry for local community development', FAO Forestry Paper No. 7.

FAO (1979). 'Forestry for rural communities'.

FAO (1980). 'Towards a strategy for development'.

FAO (1981). 'Map of the fuelwood situation in the developing countries' (with explanatory note).
FAO (1982). 'Tropical forest resources'. FAO Forestry Paper No. 30. This publication is a synthesis of three more detailed FAO/UNEP regional studies:
Forest Resources of Tropical Asia
Los Recursos Forestales de la America Tropical
Forest Resources of Tropical Africa, Vols. I and II
FAO/IBPGR (1980). 'Genetic resources of tree species in arid and semi-arid areas—a survey for the improvement of rural living in Latin America, Africa, India and Southwest Asia', FAO/IBPGR Report, Rome.
Palin, D. (1980). *Management of Development Forestry.* GCP/RAS 46 (SWE), FAO, Rome.
Singh, K. D., and Lanly, J. P. (1979). 'FAO Activities in the Field of Forest Resource Surveys'. FO.MISC/79/14, FAO, Rome.

Appendix

Names and addresses of members of the Dehra Dun Working Party

J. E. M. Arnold	Forestry Department, Food and Agriculture Organization of the United Nations, Rome, Italy
Gerardo Budowski	CATIE, Turrialba, Costa Rica
T. Burgess	North-East London Polytechnic, Romford Road, London, UK
Celso P. Dias	Socio-Economics Research Division, Forest Research Institute, Ministry of Natural Resources, College, Laguna 3720, Philippines
R. C. Ghosh	Director, Forestry Research Institute, Dehra Dun, India
Arturo Gomez-Pompa	Institute for Research on Biotic Resources, Xalapa, Veracruz, Mexico
P. N. Gupta	Uttar Pradesh Forest Department, Dehra Dun, India
E. G. Hallsworth	The IFIAS–SOS Project, Science Policy Research Unit, University of Sussex, Brighton, Sussex, UK
F. R. Honey*	CSIRO Division of Land Resources Management, Floreat Park, Western Australia
W. J. Lusigi	UNESCO Regional Office for Science and Technology for Africa, PO Box 30592, Nairobi, Kenya
Lill Lundgren	Department of Physical Geography, University of Stockholm, Sweden
D. N. Misra	Indian Institute of Forest Management, Ahmedabad, India
Ratna Murdia	Tata Institute of Social Sciences, Deonar, Bombay–400 088, India
Wanjiku Mwagiru	Department of Geography, University of Nairobi, Kenya
Ishemat Soerianegara	Director, SEAMEO Regional Center for Tropical Biology (BIOTROP), Bogor, Indonesia

R. N. Tewani*	Xavier Labour Relations Institute, Jamshedpur, India
E. S. Thangam	Forest Department, Arunachal Pradesh, India
Pisit Vorauri	Faculty of Agriculture, Chiang Mai University, Chiang Mai, Thailand
L. J. Webb	School of Australian Environmental Studies, Griffith University, Brisbane, Australia

Copies of the papers presented by the members marked with an asterisk, which are not published in this book can be obtained from the authors.

Glossary

Adivasis	A term used in parts of India for tribal people.
Godown	A storehouse.
Kaingin	The practice of 'slash and burn', where the forest is cut and burnt in small patches which are cultivated for two or more years, and then allowed to revert to forest.
Kainginero	A peasant cultivator who practices Kaingin.
Malki land	Private land, as distinct from communally owned land (India).
Magu	A kind of raw silk.
Sal	An Indian forest species — *Shorea robusta.*
Tendu	An Indian forest species — *Diopyrus melanoxylon.*
Swidden	The practice of slash and burn ≡ Kaingin.

Index

S